SPACE SCIENCE, EXPLORATION AND POLICIES

GEOSPATIAL DATA

BACKGROUND, FEDERAL AND STATE USE, AND DUPLICATION CHALLENGES

SPACE SCIENCE, EXPLORATION AND POLICIES

Additional books in this series can be found on Nova's website under the Series tab.

Additional e-books in this series can be found on Nova's website under the e-book tab.

SPACE SCIENCE, EXPLORATION AND POLICIES

GEOSPATIAL DATA

BACKGROUND, FEDERAL AND STATE USE, AND DUPLICATION CHALLENGES

MEREDITH HAYNES
EDITOR

nova publishers
New York

NOTICE TO THE READER

The Publisher has taken reasonable care in the preparation of this book, but makes no expressed or implied warranty of any kind and assumes no responsibility for any errors or omissions. No liability is assumed for incidental or consequential damages in connection with or arising out of information contained in this book. The Publisher shall not be liable for any special, consequential, or exemplary damages resulting, in whole or in part, from the readers' use of, or reliance upon, this material. Any parts of this book based on government reports are so indicated and copyright is claimed for those parts to the extent applicable to compilations of such works.

Independent verification should be sought for any data, advice or recommendations contained in this book. In addition, no responsibility is assumed by the publisher for any injury and/or damage to persons or property arising from any methods, products, instructions, ideas or otherwise contained in this publication.

This publication is designed to provide accurate and authoritative information with regard to the subject matter covered herein. It is sold with the clear understanding that the Publisher is not engaged in rendering legal or any other professional services. If legal or any other expert assistance is required, the services of a competent person should be sought. FROM A DECLARATION OF PARTICIPANTS JOINTLY ADOPTED BY A COMMITTEE OF THE AMERICAN BAR ASSOCIATION AND A COMMITTEE OF PUBLISHERS.

Additional color graphics may be available in the e-book version of this book.

Library of Congress Cataloging-in-Publication Data

ISBN: 978-1-63483-524-4

Published by Nova Science Publishers, Inc. † New York

CONTENTS

PREFACE

The federal government collects, maintains, and uses geospatial information—data linked to specific geographic locations—to help support varied missions, including national security and natural resources conservation. To coordinate geospatial activities, in 1994 the President issued an executive order to develop a National Spatial Data Infrastructure—a framework for coordination that includes standards, data themes, and a clearinghouse. This book describes the geospatial data that selected federal agencies and states use and how much is spent on geospatial data; assess progress in establishing the National Spatial Data Infrastructure; and determine whether selected federal agencies and states invest in duplicative geospatial data.

In: Geospatial Data
Editors: Meredith Haynes

ISBN: 978-1-63483-524-4

Chapter 1

GEOSPATIAL DATA: PROGRESS NEEDED ON IDENTIFYING EXPENDITURES, BUILDING AND UTILIZING A DATA INFRASTRUCTURE, AND REDUCING DUPLICATIVE EFFORTS*

United States Government Accountability Office

WHY GAO DID THIS STUDY

The federal government collects, maintains, and uses geospatial information—data linked to specific geographic locations—to help support varied missions, including national security and natural resources conservation. To coordinate geospatial activities, in 1994 the President issued an executive order to develop a National Spatial Data Infrastructure—a framework for coordination that includes standards, data themes, and a clearinghouse. GAO was asked to review federal and state coordination of geospatial data.

GAO's objectives were to (1) describe the geospatial data that selected federal agencies and states use and how much is spent on geospatial data; (2) assess progress in establishing the National Spatial Data Infrastructure; and (3) determine whether selected federal agencies and states invest in duplicative

* This is an edited, reformatted and augmented version of a United States Government Accountability Office publication, No. GAO-15-193, dated February 2015.

geospatial data. To do so, GAO identified federal and state uses of geospatial data; evaluated available cost data from 2013 to 2015; assessed FGDC's and selected agencies' efforts to establish the infrastructure; and analyzed federal and state datasets to identify duplication.

WHAT GAO RECOMMENDS

GAO suggests that Congress consider assessing statutory limitations on address data to foster progress toward a national address database. GAO also recommends that OMB improve its oversight of FGDC and federal agency initiatives, and that FGDC and selected agencies fully implement initiatives. The agencies generally agreed with the recommendations and identified plans to implement them.

WHAT GAO FOUND

Federal agencies and state governments use a variety of geospatial datasets to support their missions. For example, after Hurricane Sandy in 2012, the Federal Emergency Management Agency used geospatial data to identify 44,000 households that were damaged and inaccessible and reported that, as a result, it was able to provide expedited assistance to area residents. Federal agencies report spending billions of dollars on geospatial investments; however, the estimates are understated because agencies do not always track geospatial investments. For example, these estimates do not include billions of dollars spent on earth-observing satellites that produce volumes of geospatial data. The Federal Geographic Data Committee (FGDC) and the Office of Management and Budget (OMB) have started an initiative to have agencies identify and report annually on geospatial-related investments as part of the fiscal year 2017 budget process.

FGDC and selected federal agencies have made progress in implementing their responsibilities for the National Spatial Data Infrastructure as outlined in OMB guidance; however, critical items remain incomplete. For example, the committee established a clearinghouse for records on geospatial data, but the clearinghouse lacks an effective search capability and performance monitoring. FGDC also initiated plans and activities for coordinating with state governments on the collection of geospatial data; however, state officials

GAO contacted are generally not satisfied with the committee's efforts to coordinate with them. Among other reasons, they feel that the committee is focused on a federal perspective rather than a national one, and that state recommendations are often ignored. In addition, selected agencies have made limited progress in their own strategic planning efforts and in using the clearinghouse to register their data to ensure they do not invest in duplicative data. For example, 8 of the committee's 32 member agencies have begun to register their data on the clearinghouse, and they have registered 59 percent of the geospatial data they deemed critical. Part of the reason that agencies are not fulfilling their responsibilities is that OMB has not made it a priority to oversee these efforts. Until OMB ensures that FGDC and federal agencies fully implement their responsibilities, the vision of improving the coordination of geospatial information and reducing duplicative investments will not be fully realized.

OMB guidance calls for agencies to eliminate duplication, avoid redundant expenditures, and improve the efficiency and effectiveness of the sharing and dissemination of geospatial data. However, some data are collected multiple times by federal, state, and local entities, resulting in duplication in effort and resources. A new initiative to create a national address database could potentially result in significant savings for federal, state, and local governments. However, agencies face challenges in effectively coordinating address data collection efforts, including statutory restrictions on sharing certain federal address data. Until there is effective coordination across the National Spatial Data Infrastructure, there will continue to be duplicative efforts to obtain and maintain these data at every level of government.

ABBREVIATIONS

BLM	Bureau of Land Management
Census	U.S. Census Bureau
Commerce	Department of Commerce
DHS	Department of Homeland Security
FEMA	Federal Emergency Management Agency
FGDC	Federal Geographic Data Committee
FSA	Farm Service Agency
GIS	geographic information system
Interior	Department of the Interior
IT	information technology

LIDAR	light detection and ranging
NGAC	National Geospatial Advisory Committee
NGDA	National Geospatial Data Asset
NOAA	National Oceanic and Atmospheric Administration
NSDI	National Spatial Data Infrastructure
NSGIC	National States Geographic Information Council
OMB	Office of Management and Budget
USDA	U.S. Department of Agriculture
USGS	U.S. Geological Survey
USPS	United States Postal Service

* * *

February 12, 2015

The Honorable Thomas R. Carper
Ranking Member
Committee on Homeland Security and Governmental Affairs
United States Senate

The Honorable Orrin G. Hatch
United States Senate

The Honorable James E. Risch
United States Senate

The Honorable Mark R. Warner
United States Senate

The federal government collects, maintains, and uses geospatial data—information linked to specific geographic locations—to help in decision making and to support many functions, including national security, law enforcement, health care, environmental protection, and natural resources conservation.

Among the many activities that can depend on critical analysis of geospatial data are maintaining roads and other critical transportation infrastructures and quickly responding to natural disasters, such as floods, hurricanes, and fires.

Multiple federal agencies provide services at the same geographic locations and may independently collect similar geospatial data about those locations, thus raising the question of how well the nation's investments in geospatial data are coordinated. In an attempt to coordinate geospatial activities and discourage duplication, in 1994, the President issued an executive order to develop a National Spatial Data Infrastructure (NSDI). Consisting of data themes, standards, metadata, a clearinghouse, and partnerships, NSDI was expected to facilitate the efficient collection, sharing, and dissemination of geospatial data among all levels of government, the private sector, and the public.

However, the initiation of the NSDI did not halt duplicative investments in geospatial data. In August 2012, the Department of the Interior (Interior) estimated that the federal government invested billions of dollars in geospatial data annually, and reported that duplication among investments was common.[1] Additionally, in November 2012, we reported that federal efforts to acquire geospatial data were uncoordinated and that the government was acquiring duplicative data.[2] We made several recommendations to improve coordination among federal agencies, and the status of those recommendations is discussed later in this report.

Given your interest in reducing duplicative investments, you asked us to review federal and state efforts to improve coordination in acquiring geospatial data. Our objectives were to (1) describe the geospatial data that selected federal agencies and states are using to support their missions, and identify how much is spent on geospatial data; (2) assess progress in establishing the NSDI; and (3) determine whether selected federal agencies and states invest in duplicative geospatial data.

To address our objectives, we selected two key data types, seven federal agencies, and five states. We selected the two key data types— addresses and imagery resulting from aerial photography—based on (1) the need to appropriately scope our review due to the large amounts of geospatial data, (2) the importance of these types of data to the work of the federal and state governments, and (3) the potential for duplication within these data types, as noted by prior federal reports.

We selected agencies based on several factors, including agencies whose missions were closely tied to the selected data types (addresses and aerial photography), as well as including a mix of agencies who participate in the Office of Management and Budget's (OMB) required geospatial clearinghouse[3] and those who do not. In addition, we sought to include some agencies that were not the subject of our prior geospatial report.[4] The federal agencies we selected because of their reliance on address data are the Department of Commerce's (Commerce) U.S. Census Bureau (Census), the Department of Homeland Security's (DHS) Federal Emergency Management Agency (FEMA), and the United States Postal Service (USPS).[5] The federal agencies we selected because of their reliance on aerial photography are the U.S. Department of Agriculture's (USDA) Farm Service Agency (FSA), Commerce's National Oceanic and Atmospheric Administration (NOAA), the Department of the Interior's (Interior) Bureau of Land Management (BLM), and Interior's U.S. Geological Survey (USGS).

We selected the states in our review based on whether they had state imagery and address data programs and whether the state hosted federal imagery programs. In addition, we sought states that would provide regional diversity within the United States. The selected states are Maryland, Montana, Ohio, South Carolina, and Washington.

To describe the geospatial data that the selected federal agencies and states use to support their missions, we reviewed agency and state documentation and responses to structured questions, and conducted interviews with agency and state officials.

To assess progress in establishing the NSDI, we compared agencies' activities against guidance found in OMB documentation and an executive order regarding geospatial activities.[6] We also interviewed relevant agency officials about ongoing and planned initiatives. To determine whether the selected federal agencies and states invest in duplicative data, we compared OMB and GAO guidance on reducing duplication to practices in place on the geospatial platform as well as agency and state repositories.[7] Specifically, we analyzed federal and state collections of two types of geospatial data—addresses and aerial photography—to identify datasets that appeared to overlap. We then interviewed relevant federal and state officials to determine if these potential overlaps were examples of duplication.

We conducted this performance audit from January 2014 to February 2015 in accordance with generally accepted government auditing standards. Those standards require that we plan and perform the audit to obtain sufficient, appropriate evidence to provide a reasonable basis for our findings and conclusions based on our audit objectives. We believe that the evidence obtained provides a reasonable basis for our findings and conclusions based on our audit objectives. Details of our objectives, scope, and methodology are contained in appendix I.

Background

Geospatial data describe features or phenomena that can be referenced to specific locations relative to the earth's surface. For example, features such as buildings, rivers, road intersections, power plants, and national parks can all be identified by their locations. In addition, phenomena such as wildfires, the spread of an enterovirus, and the thinning of trees due to acid rain can also be tracked by their geographic locations.

Individuals can analyze geospatial data in geographic information systems (GIS)—systems of computer software, hardware, and data used to capture, store, manipulate, analyze, and graphically present a potentially wide array of geospatial data. The primary function of a GIS is to link multiple sets of geospatial data and display the combined information as maps with different layers of information. Assuming that all of the information is at the same scale and has been formatted according to the same geospatial standards, users can potentially overlay geospatial data about any number of specific topics to examine how the data in the various layers interrelate.

Each layer of a GIS map typically represents a single theme made up of one or more sets of data, each of which could be derived from a source completely different from the others. For example, one theme could represent all of the streets in a specific area. Another theme could correspond to all of the buildings in the same area, and others could show vegetation and water resources. Analyzing this layered information as an integrated whole can significantly aid decision makers in considering complex choices, such as where to locate a fire station building to best serve the greatest number of citizens. Figure 1 portrays the concept of data themes in a GIS.

Coordination of Geospatial Data in the Federal Government

For many years, the federal government has taken steps to coordinate geospatial activities both within and outside the federal government to discourage the duplication of data and the inefficient use of resources. In 1953, the Bureau of the Budget[8] first issued Circular A-16, encouraging expeditious surveying and mapping activities across all levels of government and avoidance of duplicative efforts. In 1990, OMB revised Circular A-16 to, among other things, establish the Federal Geographic Data Committee (FGDC) as an interagency coordinating body to promote the coordinated use, sharing, and dissemination of geospatial data nationwide. FGDC is chaired by the Secretary of the Interior and supported by a program office within USGS.

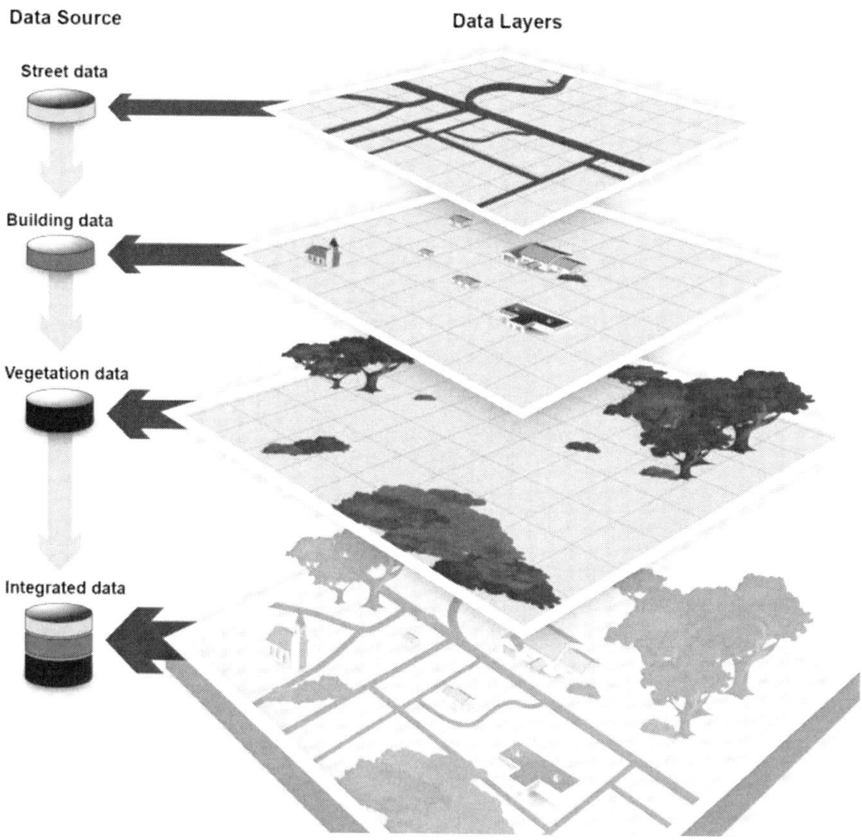

Source: GAO. | GAO-15-193.

Figure 1. Visual Representation of Data Themes in a Geographic Information System.

To strengthen that guidance, in 1994 the President issued Executive Order 12906 to address wasteful duplication and incompatibility of geospatial data, and assigned FGDC the responsibility to coordinate the development of the National Spatial Data Infrastructure (NSDI). In 2002, OMB again revised Circular A-16 to further describe the components of NSDI; clearly define agency responsibilities for acquiring, maintaining, distributing, using, and preserving geospatial data; and reaffirm FGDC's role as the interagency coordinating body for NSDI-related activities. The circular established the following five components of NSDI and described how these components were to be implemented.

- **Data themes.** Data themes are topics of national significance, such as cadastre, which includes rights and interests in real property and surveys, and land use-land cover, which includes land surface features and use. OMB Circular A-16 identified 34 data themes and identifies the lead agency or agencies for each theme. Each data theme is to consist of one or more electronic data records, known as datasets.
- **Standards.** Geospatial standards provide common and repeatable rules or guidelines for the development, documentation, and exchange of geospatial datasets.
- **Metadata.** Metadata are information about datasets, such as content, source, accuracy, method of collection, and point-of-contact. Metadata are used to facilitate the search of and access to datasets within a data library or clearinghouse, and enable potential users to determine the data's applicability for their use.
- **National Spatial Data Clearinghouse.** The clearinghouse is intended to be a centralized repository that contains geospatial metadata records from federal agencies, state and local governments, and academic and private sector organizations. Executive Order 12906 and OMB guidance require federal agencies to identify their existing and planned geospatial investments in the clearinghouse, and to search the clearinghouse for cost-saving opportunities before acquiring new geospatial data. In 2003, FGDC created the Geospatial One-Stop as the clearinghouse in order to provide "one-stop" access to geospatial metadata from a centralized database and search function. In October 2011, the Geospatial One-Stop was retired, and FGDC initiated the Geospatial Platform.[9] As of September 2014, there were approximately 88,000 geospatial metadata records registered on the Geospatial Platform, of which about 42,000 were from federal sources. Individuals can search the metadata repository and access the data through two different portals: the Geospatial Platform[10] and Data.gov.[11]
- **Partnerships.** Partnerships are efforts aimed at involving all stakeholders (e.g., federal, tribal, state, and local government, as well as academic institutions) in the development of the NSDI. Some of these partnerships include the National Geospatial Advisory

Committee (NGAC),[12] the National Digital Orthoimagery Program,[13] and the National Digital Elevation Program.[14]

In 2010, OMB provided supplemental guidance that further defined and clarified Circular A-16 and focused on managing geospatial data as a capital asset. The guidance established the concept of National Geospatial Data Assets (NGDA), which are the most significant data themes and datasets. It also encouraged agencies to adopt and implement a portfolio management approach for their investments in geospatial data. By moving to a portfolio management approach in individual agencies, the guidance sought to create a government-wide NGDA portfolio.

As of August 2014, FGDC reported that there were 16 NGDA themes and 188 NGDA datasets. Figure 2 provides an overview of the NGDA portfolio and lists the 16 NGDA data themes. (See app. II for descriptions of the 16 NGDA data themes.)

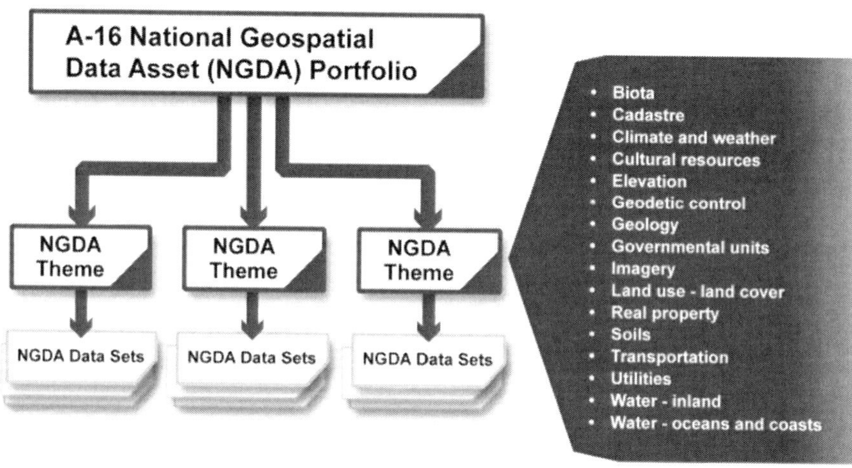

Source: GAO analysis of Federal Geographic Data Committee data. I GAO-15-193.

Figure 2. Overview of the National Geospatial Data Asset Portfolio.

To fulfill its responsibilities, FGDC is governed by a steering committee— an interagency decision-making body that provides leadership and policy direction in support of the development of the NSDI. The Secretary of the Interior chairs the committee, and the Deputy Director for Management of OMB is the Vice-Chair of the committee. All departments or agencies that are responsible for geospatial data themes or have activities in geographic information or geospatial data collection or uses are required by OMB to be members of FGDC. Thirty-two agencies are currently members of the steering committee and are represented by their senior agency officials for geospatial data.[15] These senior agency officials are responsible for overseeing, coordinating, and facilitating their respective agency's implementation of geospatial requirements, policies, and activities. FGDC is supported by an Office of the Secretariat that consists of about 10 people located in USGS who do the day-to-day work of supporting, managing, and coordinating the activities of FGDC.

In December 2007, the Secretary of the Interior created the NGAC [16] to provide the department and FGDC with advice and recommendations related to the management of federal and national geospatial programs, development of the NSDI, and the implementation of related federal guidance. Members of the committee include approximately 30 officials from federal, state, local, and tribal governments, the private sector, and academia. The advisory committee meets every 3 to 4 months, and provides a forum to convey views representative of non-federal stakeholders in the geospatial community.

OMB's Roles, Responsibilities, and Oversight

OMB has specific responsibilities for federal information technology (IT) systems and acquisition activities—including GIS—to help ensure their efficient and effective use. The three key laws that outline these responsibilities are the Paperwork Reduction Act of 1995, Clinger-Cohen Act of 1996, and E-Government Act of 2002.

- The Paperwork Reduction Act of 1995 specified OMB and agency responsibilities for managing information resources, including the management of IT. Among its provisions, this law established agency responsibility for assessing and managing the risks of major information systems initiatives. It also required that OMB develop and oversee policies, principles, standards, and guidelines for federal agency IT functions, including periodic evaluations of major information systems.
- The Clinger-Cohen Act of 1996 requires OMB to establish processes to analyze, track, and evaluate the risks and results of major capital investments in information systems made by federal agencies

and report to Congress on the net program performance benefits achieved as a result of these investments.

- The E-Government Act of 2002 establishes e-government initiatives, which encourage the use of web-based Internet applications to enhance the access to and delivery of government information and service to citizens, business partners, and employees, and among all levels of government. The act also requires OMB to report annually to certain congressional committees on the status of the e-government initiatives. In these reports, OMB is to describe the administration's use of e-government principles to improve government performance and the delivery of information and services to the public.

OMB subsequently began to fulfill the requirements established by these laws:

- In February 2002, OMB established the Federal Enterprise Architecture, which is intended to facilitate government-wide improvement through cross-agency analysis and identification of duplicative investments, gaps, and opportunities for collaboration, interoperability, and integration within and across agency programs. The Federal Enterprise Architecture is composed of five "reference models" describing the federal government's (1) business (or mission) processes and functions, independent of the agencies that perform them; (2) performance goals and outcome measures; (3) means of service delivery; (4) information and data definitions; and (5) technology standards.
- In March 2004, OMB established multiple "Line of Business" initiatives to consolidate redundant IT investments and business processes across the federal government. Later, in March 2006, OMB established the Geospatial Line of Business. Each Line of Business initiative is led by an individual agency and supported by other relevant agencies. Interior is the managing partner for the Geospatial Line of Business and the FGDC Secretariat provides project management support.

In carrying out its responsibilities, OMB uses several data collection mechanisms to oversee federal IT spending during the annual budget formulation process. Specifically, OMB requires federal departments and agencies to provide information on their IT-related investments (called exhibit 53s) and capital asset plans and business cases (called exhibit 300s).

- Exhibit 53. The purpose of the exhibit 53 is to identify all IT investments—both major and non-major—and their associated costs within a federal organization.[17] Information included in agency exhibit 53s is designed, in part, to help OMB better understand agencies' spending on IT investments. OMB guidance on formulating budgets for fiscal years 2013 and 2014 instructed agencies to identify their geospatial investments in the exhibit 53 using Federal Enterprise Architecture codes for specific functions (e.g., geospatial services, financial management, and acquisition management).
- Exhibit 300. The purpose of the exhibit 300 is to provide a business case for each major IT investment and to allow OMB to monitor IT investments once they are funded. Agencies are required to provide information on each major investment's cost, schedule, and performance.

Coordination of Geospatial Data in the State Governments

State governments rely on geospatial data to implement a wide range of responsibilities, including emergency management, taxation, wildlife management, and maintaining and managing state roads and bridges, among others. In order to support the NSDI, the National States Geographic Information Council (NSGIC) promotes geospatial coordination activities in all states. The council also advocates for states in national geospatial policy and initiatives. Members of NSGIC include state geographic information system managers, representatives from federal agencies, local government, the private sector, academia, and other professional organizations. NSGIC conducts semi-annual conferences to bring together its diverse membership. In addition, through its committees and work groups, NSGIC also develops proposals and recommendations through its advocacy efforts, including proposing a national address database and improved elevation data coordination.

NSGIC is also member of the Coalition of Geospatial Organizations, which provides a forum for organizations concerned with national geospatial issues in order to improve communications, provide educational information on relevant issues, align and strengthen policy agendas, and facilitate development of strategies to address national issues.

In addition to state-sponsored organizations, state governments typically have representation on many of the federal coordination groups, such as the National Digital Orthoimagery Program and the National Digital Elevation Program (both through NSGIC). In addition, representatives from state and local governments are members of NGAC—the committee that advises FGDC. Non-federal organizations, including stakeholders from organizations that represent state and local governments, may also serve as collaborating partners with the FGDC. These organizations include the Association of American Geographers,[18] NSGIC, the National Association of State Chief Information Officers,[19] the National Association of Counties,[20] the Western Governors' Association,[21] and the National League of Cities.[22]

Overview of Two Key Data Types: Addresses and Aerial Imagery

While federal, state, and other entities regularly utilize a wide range of geospatial data, two key types of data are addresses and aerial photography. Addresses are the location identifiers most widely used by state and local government and the public. In most cases, local or county address authorities create new addresses. From there, addresses are aggregated by several types of entities. For example, some states aggregate local addresses in order to create a statewide database to help the state carry out its mission. In addition, USPS works with local address authorities to update their address databases on an ongoing basis. USPS produces a separate address file, which is provided to others for a fee. Private sector vendors typically use this address file to create their own address data products that they sell to their customers, which include federal agencies, state and local governments, and other private sector companies. Census uses several sources to update the address file it uses to conduct its various censuses. These sources include addresses from USPS, state and local government, and other federal agencies.

To increase their usefulness, address data are often linked to the exact geographical coordinates of a specific location through a process called geocoding or geo-referencing. Geocoded addresses provide a much greater level of accuracy than can be derived from using street centerlines and address ranges. Organizations may have different requirements for what location should be geocoded. For example, emergency officials need the geocode of the home or housing structure, whereas USPS is more interested in the location of the mailbox. Figure 3 shows the difference from address points that are based on the center of the land parcels and address points that are based on the street centerline.

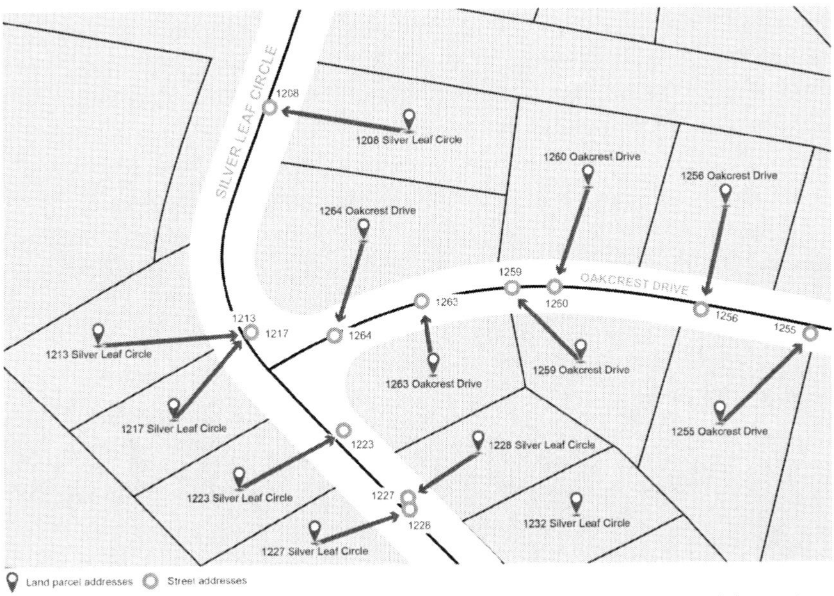

Source: GAO analysis of report prepared for Census; NAVTEQ (street addresses) and CoreLogic (land parcel addresses). | GAO-15-193.

Note: This image is for illustrative purposes only and does not reflect present-day status.

Figure 3. Difference between Address Points Based on Land Parcels and Street Centerlines.

Address data are instrumental to multiple federal, state, local, tribal, and private organizations to conduct their missions. For example, addresses are used to register voters, provide disaster relief, collect sales tax, deliver mail and packages, track diseases, and provide utilities. In addition, addresses are the most commonly used way to communicate the location of an emergency; lives and property may be lost if first responders cannot quickly obtain an address to accurately locate an emergency event. For example, in the wake of Hurricane Katrina, the lack of address information slowed and frustrated rescue and recovery operations.[23]

In addition to addresses, aerial photography is a key data type used by many levels of the government.[24] Aerial photography, which is part of the imagery data theme, records the ever-changing natural and man-made features on the earth's surface using aircraft (including planes, helicopters, and unmanned aerial systems) that are equipped with mapping cameras. When aerial photography began, imagery was acquired using film cameras. With recent advances in technology, more acquisition is being done with digital sensors. Digital sensors also offer capabilities that allow for the enhancement of the images to compensate for adverse conditions, such as shadows. This allows data collection in areas that might be obscured on film. Regardless of whether the images are collected using film or digitally, the resulting images are inspected using automated and visual methods to ensure accuracy. Orthoimagery, which are images of the earth's surface that have been corrected to resolve the displacement of features in the photo due to camera angles and terrain relief, can also be used as maps.

Source: Google, Inc. | GAO-15-193.

Figure 4. Leaf-on Aerial Photography.

Source: Maryland Department of Information Technology. | GAO-15-193.

Figure 5. Leaf-off Aerial Photography.

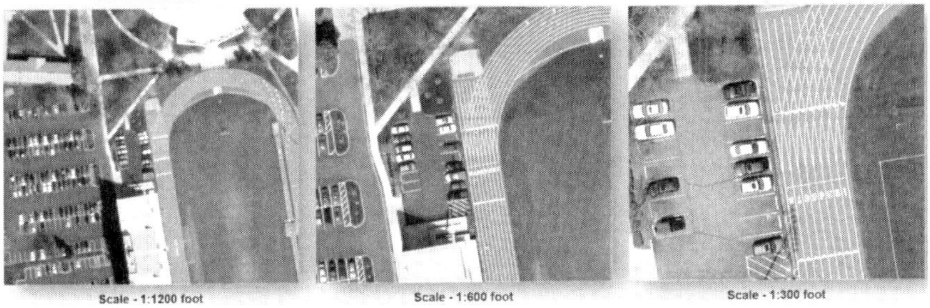

Source: Ohio Statewide Imagery Program. | GAO-15-193.

Figure 6. Example of Aerial Photography at Different Resolutions.

Most aerial photography is acquired through private sector vendors and contractors. Aerial photography can be purchased from vendors that maintain libraries of existing images, or contracted for specifically according to a customer's need for a particular product. When contracting for aerial photography collections, many elements need to be considered, including the size of the area to be photographed, the terrain (urban or non-urban, mountains or plains), vegetation cover (agricultural or forest, leaf-on or leaf-off[25]), sunlight (time of day or time of year) and angle (straight down—known as vertical, or an at angle—known as oblique). Most imagery is acquired as a four-band product, which can be viewed as either a natural color or color infrared image. See figures 4 and 5 below for a comparison of leaf-on and leaf-off imagery.

One important consideration in using an image is determining the level of detail necessary. Typically, the greater the level of detail required, the higher the resolution of photography necessary. Digital aerial photography resolution is generally expressed as the amount of area covered by one pixel. For example, imagery with a 1-foot resolution will cover 1 foot of ground in each pixel; this is considered relatively high resolution. Imagery with a 1-meter resolution will cover 1 meter of ground in each pixel; this is considered relatively lower resolution. See figure 6 for examples of different aerial photography resolutions.

GAO Has Previously Reported on Geospatial Data

In November 2012, we reported that while the President and OMB had established policies and procedures for coordinating investments in geospatial data, government-wide committees and selected federal departments and agencies had not effectively implemented them.[26] We made recommendations aimed at improving coordination and reducing duplication, including recommendations that

- FGDC develop a national strategy for coordinating geospatial investments;
- selected federal agencies (Commerce, Interior, and Transportation) follow federal guidance for managing geospatial investments; and
- OMB develop a mechanism to identify and report on geospatial investments.

Commerce and Interior agreed with our recommendations; Transportation neither agreed nor disagreed. OMB concurred with the need for improved collection of geospatial-related investments, but stated that it should only be achieved through improvements to broader reporting mechanisms for IT investments and data assets, and not by developing new and separate mechanisms specifically for geospatial-related assets. We discuss agencies' progress on implementing key recommendations later in this report. For further information on the status of our prior recommendations, see appendix III.

In addition to our work on geospatial data management, over the last 4 years, we have issued a series of reports that have identified federal programs and functional areas where unnecessary duplication, overlap, or fragmentation exists;[27] the actions needed to address such conditions; and the potential financial and other benefits of doing so.[28] In particular, we identified opportunities to reduce duplication and the cost of government operations in several critical IT areas. In our 2013 annual report on duplication, we reported that better coordination among federal agencies that collect, maintain, and use geospatial data could help reduce duplication of geospatial investments and provide the opportunity for potential savings of millions of dollars. In that report, we reiterated the need for action among several federal agencies, FGDC, and OMB.

FEDERAL AGENCIES AND STATES USE A WIDE VARIETY OF GEOSPATIAL DATA TO SUPPORT THEIR MISSIONS, BUT THE REPORTED COSTS ARE UNDERSTATED

Federal agencies and state governments use multiple geospatial datasets to support their varying missions and provide vital services to their constituents. Federal sources have estimated that over 80 percent of the data federal agencies produce has a geospatial component. Each of the states and agencies we reviewed uses some or all of the 16 NGDA data themes as well as address data. Further, many of the geospatial data products used by agencies and the federal government use a combination of different datasets from different themes to derive products to support their mission needs. For a list of data themes used by selected federal agencies and states, see appendix IV.

Although federal agencies and states often use geospatial data, the costs reported for procuring, updating, and using the data are understated. To address this issue, FGDC and OMB started an initiative to have federal agencies identify and report annually on geospatial-related investments as part of the fiscal year 2017 budget process.

Federal Agencies Use a Wide Range of Geospatial Data to Support Their Missions

Federal agencies reported using many of the 16 NGDA data themes, as well as address data, to support various parts of their missions. The uses of the data range from statistical assessments of the nation, to ensuring public safety, to providing accurate elevation data for national maps, to assessing agricultural health and farm land boundaries. For example, Census is responsible for conducting various statistical censuses and surveys of the American people as part of its mission. Census uses vast amounts of geospatial data, including addresses, cadastre, governmental units, and related information (such as in the transportation and water themes). Figure 7 illustrates one of Census's products, a comparison of nationwide population change from 2000 to 2010 by census tract.[29]

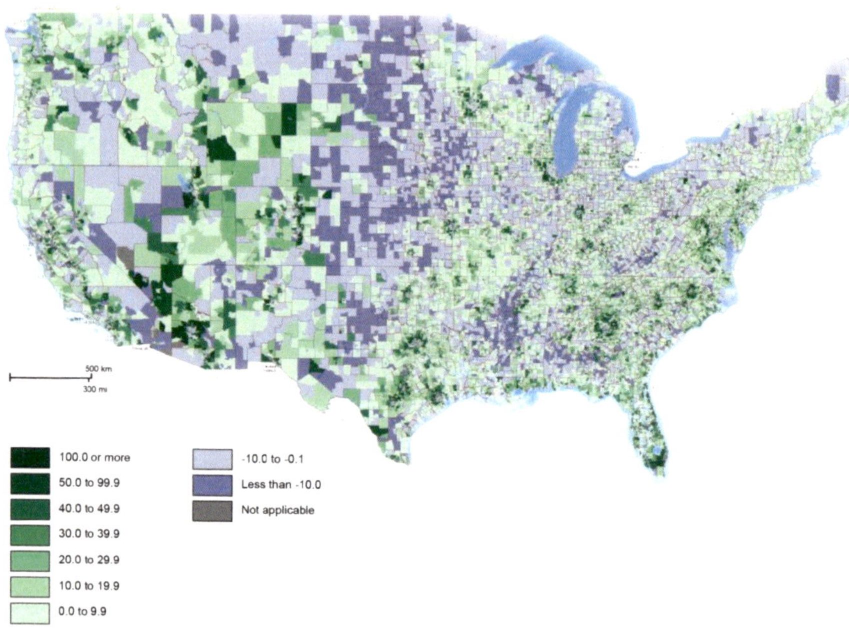

Source: Census.gov. | GAO-15-193.

Figure 7. Percent Change in Population by Census Tract from 2000 to 2010.

In addition, FEMA uses a broad range of geospatial data, including addresses, cadastre, climate and weather, imagery, and land use-land cover data to support its mission to lead Americans to prepare for, prevent, respond to, and recover from disasters. For example, during and after Hurricane Sandy in 2012, FEMA's modeling task force conducted analysis of the storm's impact on communities. Based on storm modeling data combined with real-time data coming from USGS stream and high-water markers, NOAA weather analysis and aerial imagery, and other sources, the task force was able to quickly identify 44,000 households that were both damaged and inaccessible, and as a result, FEMA reported that it was able to provide over $130 million in expedited assistance to area residents. Figure 8 illustrates houses that the task force was able to review to make determinations about their livability— even though the houses were inaccessible.

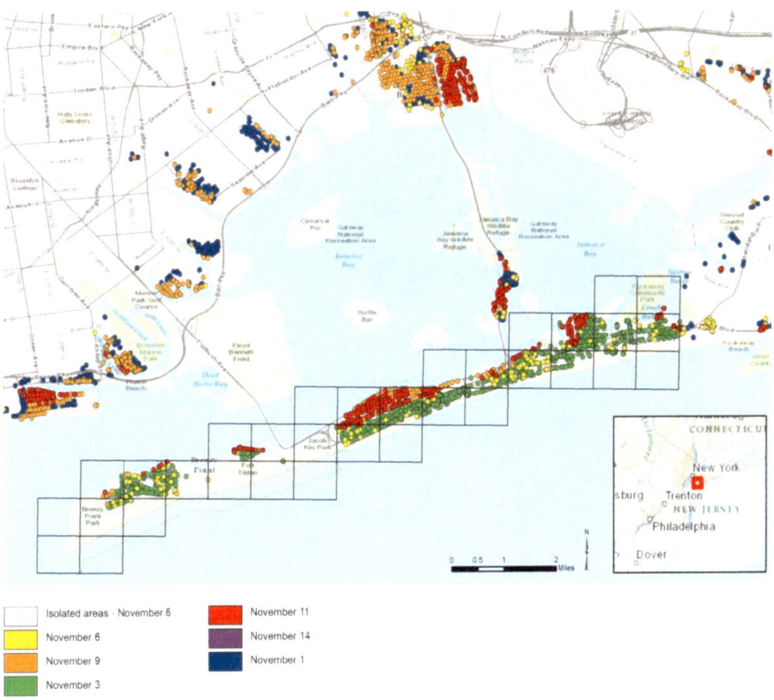

Source: Federal Emergency Management Agency. | GAO-15-193.

Figure 8. Inaccessible Locations in Jamaica Bay Region, New York (by ZIP Code + 4).

Further, one of USGS's geospatial programs is the National Elevation Dataset, which collects publicly available elevation data into a central location. The dataset serves as the elevation layer of The National Map.[30] Elevation data are used for a wide variety of purposes, including—just within USGS—natural resources management, water quality mapping, coastal zone studies, flood risk management, and determining geologic hazards. One of the ways that elevation information is collected is through the use of LIDAR (or light detection and ranging), conducted using airplane flyovers. Figure 9 shows a digital rendering of the elevation of Mount St. Helens in Washington State.

In addition, the USDA's Farm Service Agency manages the National Agriculture Imagery Program. The program acquires 1-meter resolution aerial imagery of the continental United States during the agricultural growing season.

The program's current goal is to acquire imagery of the entire contiguous United States on an annual basis; however, agency officials noted that current funding levels only allow them to collect for the entire country on a 2-year cycle. This leaf-on imagery is used by a wide range of people to improve their understanding of land cover and land use nationwide. The program's imagery serves as the base layer of many GIS programs and helps maintain FSA crop boundaries, as well as administer farm records, commodity, conservation, disaster, and lending programs. Figure 10 provides an example of the imagery collected.

Source: Geological Survey. | GAO-15-193.

Figure 9. Mount St. Helens, Washington.

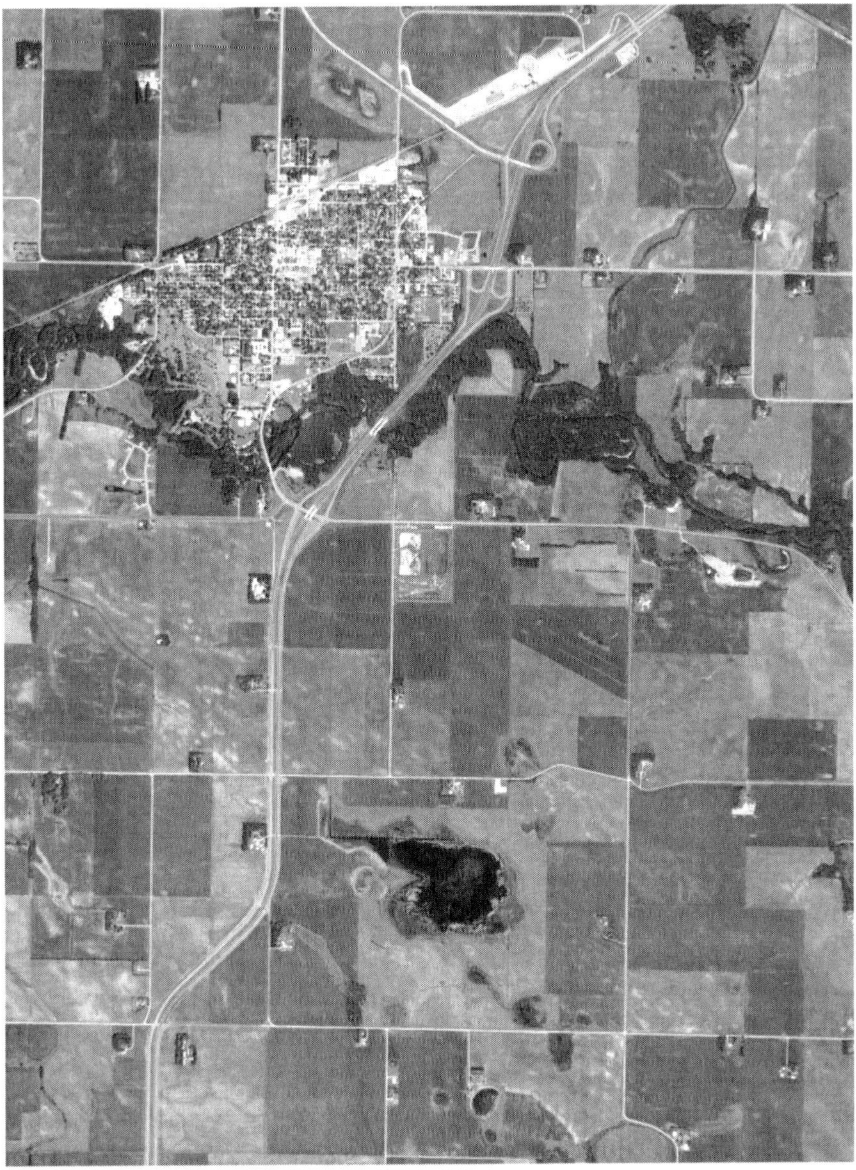

Source: National Agriculture Imagery Program. | GAO-15-193.

Figure 10. Image of Madelia, Minnesota, and Surrounding Farms during 2013 Growing Season.

States Use a Wide Variety of Geospatial Data to Support Their Missions

Selected states reported using many—and in numerous cases, all—of the 16 NGDA data themes, as well as address data, to support various parts of their missions. These uses include emergency management, focusing public health initiatives, and managing tax rates and revenues. For example, one of the ways in which Ohio uses extensive geospatial data is a site-specific state address database that uses data aggregated from local governments to assist emergency personnel, among other uses. An Ohio official noted that the response system has already saved lives because the range of addresses associated with a stretch of land is not always well coordinated with the way the roads work. Figure 11 depicts an example where address ranges that appeared to be contiguous were not contiguous because the road abruptly ended before all of the addresses were reached. With data in the improved response system, state officials noted that emergency responders could avoid a 4-mile trip around neighboring roadways, leading to a 4-minute-faster response time.

Source: State of Ohio, Preble Country. | GAO-15-193.

Figure 11. Example of Street Discontinuity, Preble County, Ohio.

Additionally, in South Carolina, a university-sponsored Medicaid policy research group[31] works with the state to use geospatial data to improve delivery of medical services to Medicaid-eligible populations. Because of a documented connection between higher rates of infectious and chronic diseases in socioeconomically deprived areas of the state, the research group developed an index that measures three key components (the percentage of persons without a high school diploma, the percentage of persons living below the poverty level, and the percentage of households with no vehicle), along with urban and rural populations measures, and layers this information on a map of the state. The derived product, pictured in figure 12 below, allows state officials to target resources in certain high-need geographic areas. According to the policy group, such tools may be especially valuable in targeting the state's limited resources to prevent, diagnose, and manage chronic conditions among Medicaid recipients.

Palmetto SADI© high deprivation areas in South Carolina[a]

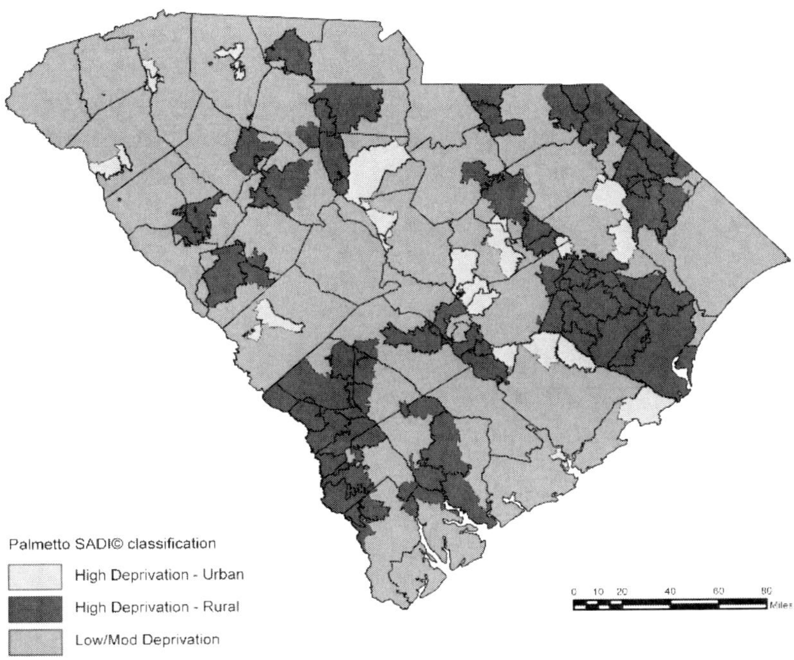

Source: University of South Carolina. Institute for Families in Society. Division of Policy and Research on Medical and Medicare. I GAO-15-193.

Note:

[a]High deprivation areas are defined as the top quartile of ZIP Code tabulation areas identified using the Palmetto Small Area Deprivation Index© (SADI) scores, 2013. ZIP Code tabulation areas are Census's generalized representations of U.S. Postal Service ZIP Codes, created by aggregating census blocks. The Palmetto SADI scores measure three components (the percentage of persons without a high school diploma, the percentage of persons living below the poverty level, and the percentage of households without a vehicle). Palmetto SADI© 2014, University of South Carolina. All rights reserved. Used by permission.

Figure 12. A Map of Palmetto Small Area Deprivation Index, as of October 2014.

Further, Washington State's Department of Revenue uses geospatial data to assist in determining the correct tax rates and payees for utilities taxes. Specifically, in Washington, utilities are taxed based on the localities they pass through—and there are over 3,200 tax code areas in the state. As a result, the analysis of where the utilities go—and which tax districts collect the revenue—can be highly complex and very sensitive. Recently, the department conducted an exercise to improve its understanding of these data. Figure 13 shows one example of this exercise. It overlays tax district information (blue) and a local railroad right of way (red) on top of satellite imagery to more accurately determine the tax districts the railroad passes through.

Source: Washington State Department of Revenue. | GAO-15-193.

Note: Map image is the intellectual property of Esri and is used herein under license.

Figure 13. Tax Jurisdictions in Walla Walla, Washington, and Railroad Right of Way.

Federal Agencies and States Reported Significant Spending on Geospatial Data and Systems, but the Amount Is Understated Because Not All Costs Are Tracked

Federal agencies and states reported spending significant funds on geospatial data and supporting systems, but the total amount spent is understated because not all costs are tracked. Identifying the cost of geospatial data has been an ongoing problem for the federal government. OMB and Interior have tried to estimate the amount spent on geospatial data and systems, but the estimates are either old or not comprehensive. In 2006 and 2007, OMB made two data calls directly to federal agencies to determine their spending on geospatial investments. However, according to OMB, neither of these data calls was complete or reliable— largely because agencies either provided incomplete information or did not respond at all. Although OMB's data collection was not complete, the agencies reported that they had planned to spend about $1.89 billion in geospatial data and services from fiscal year 2007 through 2009, of which about $1.53 billion, or 81 percent, was to be on geospatial data. Further, in 2012, Interior estimated, as part of its budget documentation, that the federal government was investing billions of dollars on geospatial data annually.[32] However, the department did not provide a comprehensive breakdown of that estimate.

OMB and agencies currently employ multiple mechanisms for collecting cost information that provide insights into the billions of dollars spent on geospatial data and systems. Current mechanisms for collecting data on geospatial data investments include (1) information on agency IT investments, capital asset plans, and business cases that were provided to OMB as part of the budget process; (2) data on USASpending.gov—a website that includes data on federal contract, grant, and assistance awards; and (3) individual agency and state estimates.

However, these estimates are understated because agencies do not always categorize geospatial investments in ways that allow them to be tracked. Further, we previously found that USASpending.gov award data were significantly underreported and where data were reported, they were not usually consistent with the information in agency records.[33] Each of the three mechanisms for collecting cost information, and its limitations, are discussed in more detail in the following sections.

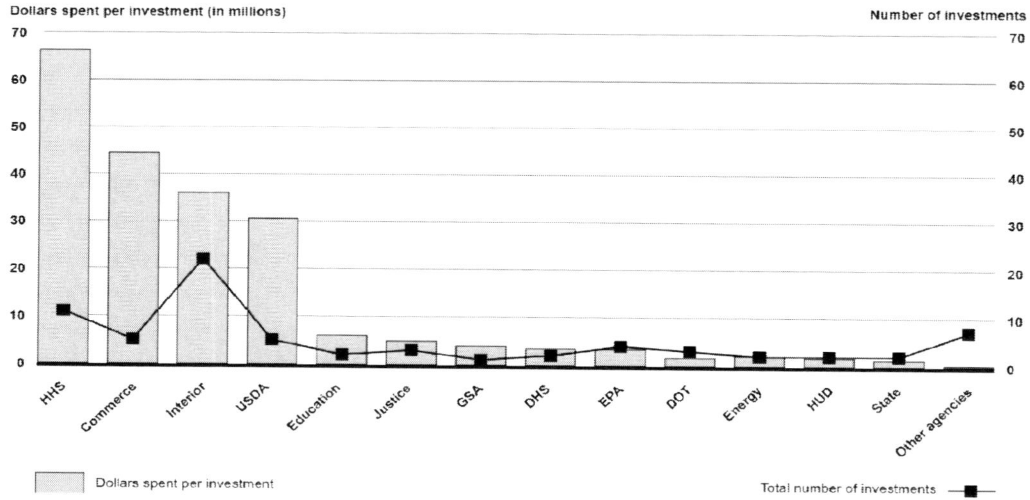

Acronyms and abbreviations:
HHS (U.S. Department of Health and Human Services), Commerce (U.S. Department of Commerce), Interior (U.S. Department of the Interior), USDA (U.S. Department of Agriculture), Education (U.S. Department of Education), Justice (U.S. Department of Justice), GSA (General Services Administration), DHS (U.S. Department of Homeland Security), EPA (U.S. Environmental Protection Agency), DOT (U.S. Department of Transportation), Energy (U.S. Department of Energy), HUD (U.S. Department of Housing and Urban Development), State (U.S. Department of State)

Source: GAO analysis of exhibit 53 data, as of August 2014. | GAO-15-193

Note: Other agencies include the Departments of Labor and the Treasury, National Science Foundation, Small Business Administration, Social Security Administration, and the U.S. Army Corps of Engineers.

Figure 14. Government-Wide Average Spending on Geospatial Information Technology Investments from Fiscal Year 2013 to 2015.

OMB Budget Data Show Billions Spent on Geospatial Investments but Do Not Capture All Costs

During the annual budget formation process, federal agencies provide OMB with information related to their IT investments (called exhibit 53s) and capital asset plans and business cases (called exhibit 300s). As part of these processes, agencies are to identify the investment using codes for specific functions (e.g., mapping or geospatial, accounting, or customer services) and provide cost information.

In their annual submissions of budget documentation, agencies categorize some, but not all, applicable investments as geospatial investments. Specifically, budget documents show that 19 agencies categorized 71 IT investments as being geospatial in nature, spending an average of $205.4 million per year for the fiscal years 2013 to 2015. Figure 14 shows which agencies reported spending the most on geospatial investments.

However, many agencies did not categorize geospatial investments as such. Specifically, 60 additional investments that were not categorized as "geospatial" are described using this term in their exhibits 53 and 300 budget documents. For example, USDA reported an IT investment called the Enterprise Spatial Mapping Service, which is described as an enterprise level, cloud-based geospatial service that supports data and GIS provisioning and consumption in a rapidly deployed portal framework. However, the investment was not reported as a geospatial investment. Further, the Department of Defense accounts for 35 of the unreported geospatial investments. These additional investments account for, on average, another $180.3 million spent yearly or planned to be spent from fiscal year 2013 to 2015 government-wide. OMB officials explained that these systems may not be categorized as geospatial because it gives agencies significant leeway in deciding which categories they select as relevant. Table 1 provides the number and anticipated cost of investments that were described as being geospatial in nature, but not specifically categorized as geospatial investments.

With the addition of the investments that were not categorized as geospatial, there are a total of 131 geospatial investments with an average of $385.7 million spent on these investments annually for the period from fiscal year 2013 through fiscal year 2015. The geospatial investments not categorized as geospatial account for just under half of all of the geospatial investments. Figure 15 provides more details on the government-wide geospatial investments.

Table 1. Number and Cost of Investments Associated with Geospatial Data That Are Not Categorized as Geospatial, Fiscal Years 2013 to 2015

Agency	Geospatial investments not categorized as geospatial	Average amount spent or planned from fiscal years 2013 to 2015 (in millions)
Department of Agriculture	5	$2.8
Department of Commerce	3	$34.3
Department of Defense	35	$32.3
Department of Energy	1	$0.05
Department of Health and Human Services	1	$13.0
Department of Homeland Security	6	$87.0
Department of the Interior	1	$3.1
Department of Transportation	5	$5.6
General Services Administration	1	$0.05
National Aeronautics and Space Administration	1	$0.1
U.S. Army Corps of Engineers	1	$2.0
Total	60	$180.3

Source: GAO analysis of exhibit 53 and 300 data, as of August 2014. | GAO-15-193

Note: The IT Dashboard does not collect or display any classified or national security-sensitive information, and thus the totals listed in the table do not reflect any classified or national security-sensitive geospatial investments. In addition, the items in the table do not add up to the total listed due to rounding.

In addition to annual budget figures, OMB collects total estimated cost information on the largest and most critical IT investments. In documents prepared for the 2015 budget, 12 IT investments were described using the word “geospatial” in their exhibit 300s. As of August 2014, the federal government had spent an estimated $1.3 billion on these investments alone. See table 2 for details on the investments.

However, similar to other OMB data collection activities, this list is not comprehensive due to limitations in what data are collected. For example, NOAA has two major weather satellite programs totaling over $22 billion in expected life-cycle costs (with over $9 billion spent to date) that are to provide critical geo-located weather and climate data well into the next decade. Because the satellites are not identified as IT investments, they were not captured in the $1.3 billion estimate. Similar to other agencies, NOAA does not identify its space systems as IT investments because OMB does not enforce this requirement.[34]

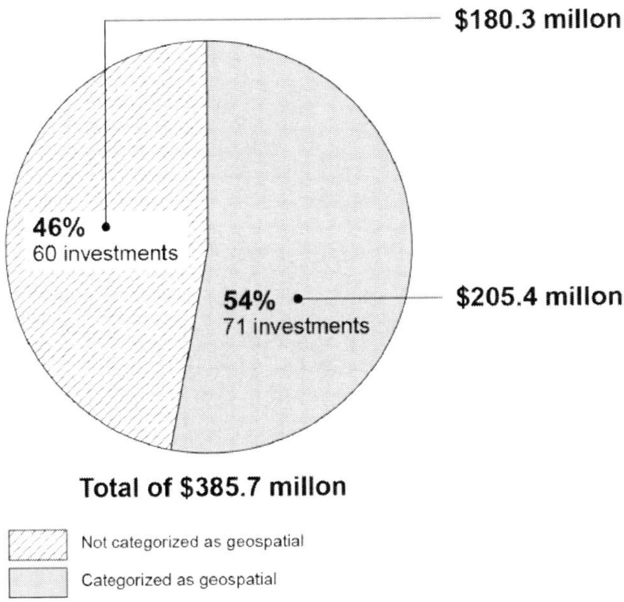

Source: GAO analysis of exhibit 53 and 300 data, as of August 2014. | GAO-15-193.

Figure 15. Government-Wide Investments Associated with Geospatial Data, Fiscal Years 2013-2015.

USASpending.gov Shows That Federal Agencies Spend Hundreds of Millions Each Year on Contracts for Geospatial Data and Services, but the Website Does Not Capture All Costs

In addition to reporting on individual IT investments to OMB, federal agencies are required by law[35] to report specific information on contract and assistance awards to USASpending.gov, a website designed to provide transparency into where the government is spending money.

While agencies used a variety of categories and descriptions, as of September 2014, federal agencies reported $493.2 million in contracting costs for geospatial data in fiscal year 2013. The Department of Defense reported spending the most on geospatial-related contract awards, with a total of $328.8 million spent. Figure 16 shows contracting costs on geospatial investments in fiscal year 2013.

Table 2. Costs of 12 Major Investments Associated with Geospatial Data, from Investment Start to August 2014

Agency	Investment	Cost in millions
Department of Commerce	Census's Geographic Support Systems	$362.5
	NOAA's Comprehensive Large Array-data Stewardship System	$243.8
	National Telecommunications and Information Administration's Federal Spectrum Management System	$46.4
Department of Defense	Air Force's Integrated Strategic Planning and Analysis Network	$75.4
Department of Health and Human Services	Food and Drug Administration Office of Regulatory Affairs Regulatory Business Information Services	$107.5
Department of Homeland Security	Customs and Border Protection's Integrated Fixed Towers	$261.0
	Immigration and Customs Enforcement's Enforcement Information Sharing	$77.1
Department of the Interior	BLM's IT Support for Resources and Mineral Land Use Planning	$34.8
	USGS's Geospatial Platform and Shared Services	$7.6
	USGS's The National Map Reengineering Project	$98.4
Department of Transportation	Federal Highway Administration's Fiscal Management Information System 5.0	$6.7
	Pipeline and Hazardous Materials Safety Administration's National Pipeline Information Exchange	$5.8
Total		**$1,327.0**

Source: GAO analysis of exhibit 300 data, as of August 2014. | GAO-15-193

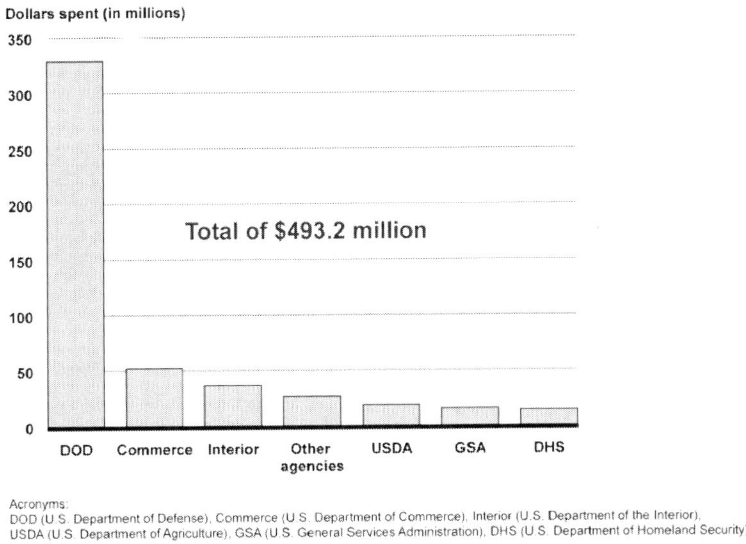

Note: Other agencies include the Departments of Health and Human Services, Housing and Urban Development, Energy, Justice, State, the Treasury, Transportation, Veterans Affairs, the Environmental Protection Agency, Federal Communications Commission, Federal Energy Regulatory Commission, National Aeronautics and Space Administration, Small Business Administration, Smithsonian Institution, and U.S. Agency for International Development.

Figure 16. Contracting Costs for Geospatial Investments in Fiscal Year 2013.

However, we recently found that USASpending.gov award data were significantly underreported and there was little consistency between the website and agency records.[36] Further, USASpending.gov data do not give a complete picture of the cost of geospatial data because there is not a definitive way to categorize the contracts as being geospatial in nature.

Agencies and States Had Difficulty Providing Comprehensive Cost Data

Selected agencies and states had difficulty providing consistent and complete cost data for their investments in geospatial data. Of seven selected agencies, one agency (FSA) was able to provide a comprehensive accounting of the cost of its geospatial data, while five (BLM, Census, FEMA, USGS, and USPS) provided estimates for selected categories of geospatial data, and one (NOAA) was unable to provide cost data even though the vast majority of its mission involves geospatial data. Agency estimates ranged from no spending on geospatial data (for data that are provided to the agency at no cost) to $26.9

million in fiscal year 2014 for the operation and maintenance of an address system. Table 3 provides agency estimates for spending on geospatial data.

However, the estimates provided often vary widely from what is reported in other data collection initiatives. For example, even though Census officials could not provide an estimate for over half of its geospatial data costs, Commerce's Office of the Inspector General previously reported that Census spent nearly $1.4 billion over a 10-year period to produce and verify address data it used in the 2010 Census.[37] The office also reported that the bureau is spending $674 million on the Geographic Support System Initiative—a multiyear effort to, among other things, improve address data quality.[38] Further, while NOAA officials could not provide an estimate of its geospatial data costs, we recently found that NOAA has spent an average of about $430 million on the operation and management of its observing systems that collect data on specific environmental conditions, such as sea surface temperature or wave height.[39]

All five selected states were able to provide estimates on some of the costs of geospatial data. These ranged from one state purchasing imagery for $130 in 2013 to another state that spends about $34 million yearly on real property data. While most states reported using nearly all the categories of geospatial data, most states were not able to provide comprehensive estimates on the cost of those data. For example, while officials from Washington estimated that the state spent approximately $3.1 million on geospatial software in 2013, they could not provide an estimate of how much was spent on other individual data categories, such as addresses or transportation. A state official attributed this to the state's decentralized governance model. Similarly, an official from Ohio was able to provide estimates for its cost of address and imagery data, but reported that the costs for the other data categories were unknown. See table 4 for more information.

We previously reported on the lack of consistent and comprehensive information about how much is spent on geospatial data, and OMB and FGDC have made plans to improve cost tracking. In November 2012, we found that OMB lacked the necessary information to accurately quantify the amount of federal dollars spent on geospatial data.[40] We recommended that OMB develop the ability to identify and report annually on all geospatial-related investments in order to improve its oversight and minimize duplication.

Table 3. Selected Agencies' Estimated Expenditures for Geospatial Data

Data category	BLM	Census	FEMA	FSA	NOAA	USGS	USPS
Addresses[a]	○	○	$722,607	n/a	n/a	n/a	$6.6 million $26.3 million[b] and $26.9 million[c] on operations and maintenance of address system
Biota	○	n/a	n/a	n/a	○	$157,000[c] for management of three datasets	n/a
Cadastre	$1 million[b] $900,000[c]	$0	$1.2 million	$0	n/a	○	n/a
Climate and Weather	$0	n/a	$0	n/a	○	○	n/a
Cultural Resources	$200,000[b] $201,000[c]	○	n/a	n/a	n/a	○	n/a
Elevation	$0	n/a	$689,976[b] $4,636[c,d]	$0	○	$13.9 million[c, e] $1.1 million for Hurricane Sandy[c]	n/a
Geodetic Control	n/a	n/a	○	n/a	○	○	n/a
Geology	$0	n/a	○	n/a	n/a	$1 million/yr for geologic map data	n/a

Table 3. (Continued)

Governmental Units, and Administrative and Statistical Boundaries	○	○	$0	$0	○	○	n/a
Imagery	$406,000 for National Agriculture Imagery Program[b,c] $1.3 million for other projects[b]	○	$287,536[b]	$9.6 million[b] $10.1 million[c]	○	$5.1 million[b] $4 million[c] $957,000/yr for National Agriculture Imagery Program	○
Land Use-Land Cover	$600,000/yr $2.5 million/yr	$0	$606,000	n/a	○	○	n/a
Real Property	○	n/a	$0	n/a	n/a	n/a	n/a
Soils	$0	n/a	n/a	$0	n/a	○	n/a
Transportation	○	○	$0	n/a	○	$84,621[c]	○
Utilities	$25,000/yr (approx.)	n/a	$0	n/a	n/a	n/a	n/a
Waters-Inland	$400,000/yr	$0	○	n/a	○	$5.7 million[c,f]	n/a
Water-Oceans and Coasts	$0	$0	$0	n/a	○	n/a	n/a

Key: ○ = Agency could not provide a cost estimate, despite reporting that it uses geospatial data from this data theme.

n/a = Agency reported that it did not use geospatial data from this data theme. Source: GAO analysis of agency reported data and documentation. | GAO-15-193

Notes:

[a]Addresses are not currently an NGDA data theme.

[b]Expenditure is from fiscal year 2013.
[c]Expenditure is from fiscal year 2014.
[d]Expenditure is partial, as fiscal year was not complete when estimate was provided.
[e]Funds contributed by other agencies to USGS for this theme equaled $23.5 million in 2014.
[f] Funds contributed by other agencies to USGS for this theme equaled $1.4 million in 2014.

Acronyms and abbreviations: BLM = Bureau of Land Management; Census = U.S. Census Bureau; FEMA = Federal Emergency Management Agency; FSA = Farm Service Agency; NOAA = National Oceanic and Atmospheric Administration; USGS = U.S. Geological Survey; USPS = United States Postal Service.

Table 4. Selected States' Estimated Expenditures for Geospatial Data for Fiscal Years 2013 and 2014

Data category	Maryland	Montana	Ohio	South Carolina	Washington
Addresses[a]	$75,000[b,c]	$80,000/yr	$450,000[b,c]	○	○
Biota	$0[d]	○	○	○	○
Cadastre	$700,000[b,c]	$800,000/yr (approx.)	○	○	○
Climate and Weather	$0[d]	$200,000/yr	○	○	○
Cultural Resources	$0[d]	○	○	○	○
Elevation	$120,000[b] $655,200[c]	○	○	○	○
Geodetic Control	$0[d]	$100,000/yr	○	○	○
Geology	$0[d]	○	○	○	○
Government Units, and Administrative and Statistical Boundaries	$0[d]	$80,000/yr	○	○	○

Table 4. (Continued)

Imagery	$810,000[c]	$130[b]	$452,000[b] $418,000[c]	$1.8 million	○
Land Use-Land Cover	$0[d]	$200,000/yr (approx.)	○	○	○
Real Property	$34.5 million[b] $35 million[c] $700,000/yr for georeferencing	○	○	○	○
Soils	$0[d]	○	○	○	○
Transportation	$2.3 million[b,c]	$35,000	○	○	○
Utilities	$0[d]	○	○	○	○
Water-Inland	$0[d]	$120,000	○	○	○
Water-Cceans and Coasts	$0[d]	n/a	○	○	○
Geospatial Software Costs	n/a	n/a	n/a	n/a	$3.1 mi lion

Key: ○= State could not provide a cost estimate, despite reporting that it uses this data theme. n/a = State reported that it did not use geospatial data from this data theme.

Source: GAO analysis of state documentation. | GAO-15-193.

Note:

[a]Addresses are not currently an NGDA data theme.

[b]Expenditure is from fiscal year 2013.

[c]Expenditure is from fiscal year 2014.

[d]Expenditure was from a year prior to fiscal years 2013 and 2014.

OMB officials generally concurred with the recommendation and, according to OMB officials, this recommendation is being addressed through FGDC's *National Geospatial Data Asset (NGDA) Management Plan*,[41] which was issued in March 2014. The plan acknowledges that it is difficult to make a determination of the total costs for federal geospatial data due to the ubiquitous nature of geospatial data (that the investment may be reflected across multiple budget codes) and the inconsistency in the use of budget codes across the government. The plan also notes that different definitions of a geospatial investment are being used by OMB, the FGDC community, and individual agencies.

Accordingly, one of the objectives of the plan is to develop and apply a standard definition of a geospatial investment in order to facilitate reporting on budgeted geospatial data investments. The plan contains two supporting actions for this objective. The first action, to have FGDC and OMB establish a standard definition of a geospatial investment for use in the budgeting process, is planned to be completed by September 2015. The second action is for federal departments and agencies to apply this definition in their respective budgeting processes. Since the fiscal year 2016 budget process is already well under way, the soonest agencies could implement the new guidance is in their fiscal year 2017 budgeting process.

A recently passed law can provide OMB with another opportunity to improve information about federal spending on geospatial data and investments. Specifically, on May 9, 2014, the President signed into law the Digital Accountability and Transparency Act of 2014,[42] which among other things, requires (1) the establishment of government-wide data standards by May 2015, (2) disclosure of direct federal spending with certain exceptions, (3) agencies to comply with the new data standards, and (4) inspectors general audits of the quality of the data made available to the public. The Comptroller General recently testified that effective implementation of the act would help promote transparency to the public and address ongoing government management challenges by expanding the quality and availability of federal spending data.[43] Until OMB and the departments have complete and reliable information on the scope, cost, and nature of investments in geospatial data, the possibility remains that the government is investing in duplicative or wasteful geospatial investments.

Table 5. National Spatial Data Infrastructure Responsibilities and Related GAO Recommendations for the Federal Geographic Data Committee

Initiative	Description and related GAO recommendation
Strategic plan	The Federal Geographic Data Committee (FGDC) is to prepare and maintain a strategic plan for the development and implementation of the National Spatial Data Infrastructure (including data themes, standards, metadata, clearinghouse, and partnerships). **GAO recommendation**: In 2012, we recommended that FGDC establish a time frame for creating and updating the strategic plan, and create and implement the plan within the established time frame. We also recommended that the plan address foundational elements of strategic planning as recognized by federal statute and OMB guidance, such as goals, objectives, and performance measures.
Portfolio management of data themes	FGDC is to manage the data themes and their associated key datasets as capital assets using a portfolio management approach, and provide guidance to federal departments on how to implement this management approach internally within their agencies. **GAO recommendation:** In 2012, we recommended that FGDC establish a time frame for completing a plan to facilitate the implementation of OMB's portfolio management guidance, and develop and implement the plan within the established time frame. We also recommended that the plan, at a minimum, include goals and performance measures, and that FGDC report annually to OMB on the progress made on efforts to improve coordination and reduce duplication among themes.
Clearinghouse	FGDC is to • develop a clearinghouse to serve as a centralized geospatial metadata repository that contains geospatial metadata records from federal agencies, state and local governments, and academic and private sector organizations; • provide a search function for the clearinghouse that permits searching of all files intended for public use, display search results in order of relevancy to search criteria; and • use analytics and customer service measurement tools to measure, analyze, and report on the effectiveness of the clearinghouse. **GAO recommendation:** In 2012, we recommended that FGDC develop guidance for federal departments on identifying planned geospatial investments on the clearinghouse.
Partnerships	FGDC is to promote and guide cooperation among federal, state, and local government agencies in the collection, production, sharing and use of geospatial information and the implementation of the National Spatial Data Infrastructure.

Source: GAO analysis of Executive Order 12906, OMB documents, and GAO-13-94.| GAO-15-193

Table 6. National Spatial Data Infrastructure Responsibilities and Related GAO Recommendations for Federal Departments

Initiative	Description and related GAO recommendation
Policy on geospatial metadata	Federal departments are to develop and implement a policy that requires their agencies to make their geospatial metadata available on the clearinghouse. **GAO recommendation:** In 2012, we recommended that the three federal departments in our prior review (Commerce, Interior, and Transportation) develop such a policy.
Procedures for accessing the clearinghouse	Federal departments are to adopt internal procedures to ensure that their agencies access the clearinghouse before they expend funds to collect or produce new geospatial data to determine (1) whether the information has already been collected by others, or (2) whether cooperative efforts to obtain the data are possible. **GAO recommendation:** In 2012, we recommended that the three federal departments in our prior review develop and implement such internal procedures.
Strategy	Federal departments are to prepare, maintain, publish, and implement a strategy for advancing geographic information and related geospatial activities appropriate to their mission. **GAO recommendation:** In 2012, we recommended that the three federal departments in our prior review develop and implement such strategies.
Partnerships	Federal departments are to coordinate and work in partnership with federal, state, and local government agencies, to efficiently and cost-effectively collect, integrate, maintain, disseminate, and preserve spatial data, building upon local data wherever possible.

Source: GAO analysis of Executive Order 12906, OMB documents, and GAO-13-94. | GAO-15-193

Progress Has Been Made in Establishing the National Spatial Data Infrastructure, but Shortfalls Remain

FGDC and selected federal departments and agencies have made progress in implementing their responsibilities for establishing the NSDI, but a number of shortfalls remain. OMB guidance and an executive order call for both FGDC and agencies to fulfill certain responsibilities in order to further the development of the NSDI and its components (data themes, standards, metadata, clearinghouse, and partnerships).[44] Specifically, FGDC is to establish a strategic plan to improve all five components, a portfolio management plan to facilitate management of data themes, a clearinghouse for coordinating investments, and partnerships for coordinating with stakeholders. Similarly, agencies are to establish their policies and procedures that address the use of the clearinghouse, strategic plans, and partnerships with states to collect and disseminate geospatial data.[45] In addition, we previously made recommendations, based on best practices in program implementation, to improve the likelihood of success on several of these geospatial initiatives.[46] Table 5 summarizes FGDC's responsibilities for key initiatives and our prior recommendations to improve those initiatives. Table 6 summarizes federal departments' responsibilities for key initiatives and our prior recommendations to improve those initiatives. The status of FGDC and the agencies' efforts are discussed later in this report.

FGDC Has Made Progress on the NSDI, but Shortfalls on Key Initiatives Exist, and States Are Dissatisfied with Coordination Efforts

FGDC has made progress in developing a strategic plan, implementing portfolio management practices for data themes, and establishing the clearinghouse infrastructure; however, these initiatives are not yet complete and shortfalls remain. In addition, selected state officials and the organizations that represent the states are generally dissatisfied with FGDC's efforts to coordinate with them. Details on each initiative follow:

- **Strategic plan.** In December 2013, FGDC issued a strategic plan for the NSDI that is consistent with OMB guidance and our prior recommendation. Specifically, the strategic plan includes a vision statement, three outcome-oriented goals, and nine objectives. Subsequently, in June 2014, FGDC created implementation plans for seven of the nine objectives that describe numerous supporting tasks necessary to achieve the objectives, and performance measures for monitoring the progress of the tasks. Implementation plans for the two remaining objectives are in development and are expected to be finalized in 2015. The implementation plans completed to date include 25 tasks that are to be completed in fiscal year 2014, as well as 40 tasks to be completed during fiscal years 2015 and 2016.

While most tasks in the implementation plans that were scheduled for completion in fiscal year 2014 were completed as scheduled, some were not. Specifically, 19 tasks were completed by September 2014. In addition, 3 tasks were completed by January 2015 and 3 other tasks remain to be completed. Completed tasks include a number of activities in support of the Geospatial Platform, such as the establishment of agency funding agreements for the platform, the creation of guidance for the use of the platform's Marketplace feature,[47] and the award of a support contract for platform operations. The tasks that are behind schedule include the development of templates for service level agreements for hosting data on the Geospatial Platform, and the creation of surveys and tools to assess the maturity of NGDA theme-lead administrative processes and the status of NGDA datasets. FGDC officials stated that they are working to finalize the target dates for the remaining fiscal year 2014 tasks by March 2015.

- **Portfolio management.** FGDC issued a portfolio management plan for NGDA assets that calls for strategic planning, coordinated development and resource pooling, goals and performance measures for each asset, investment selection criteria, and review of the health of the datasets. The plan also includes goals and performance measures, as well as tasks to formalize annual reporting to OMB, consistent with our recommendation.[48] The portfolio management plan includes 7 tasks to be completed in fiscal year 2014, as well as 20 tasks to be completed during fiscal years 2015, 2016, and 2017.

  While most of the tasks that were scheduled for completion by September 2014 were completed on time, some were not. Specifically, of the 7 tasks identified in the plan for completion by the end of fiscal year 2014, 5 were completed and 2 were expected to be completed by March 2015. Completed actions include items such as, establishing a coordination mechanism across the NGDA themes, identifying the initial baseline of NGDA datasets, and posting content for each of the data themes on the Geospatial Platform to encourage collaboration. The remaining 2 tasks include developing tools to assess and report on progress in developing NGDA datasets and themes.

- **Clearinghouse.** FGDC has created a centralized clearinghouse for geospatial metadata as part of the Geospatial Platform, and has issued guidance to federal departments on how to identify planned geospatial investments on it. However, the Geospatial Platform lacks important features called for in OMB guidance,[49] such as an effective search capability and performance monitoring. Specifically, officials from DHS and USDA noted that they had experienced problems finding datasets on the platform, and the managing partner for the Geospatial Platform stated that users had provided similar feedback.

  We also encountered difficulty with the search capability. In searching for the exact title of a dataset as it appears in the Geospatial Platform catalog in July 2014, we estimated that 62 percent of aerial photography datasets and 97 percent of address datasets did not appear in search results.[50] In November 2014, FGDC officials stated that they were aware of the issues with the search function on the Geospatial Platform and had taken steps to identify and resolve these issues. However, our more recent searches in November and December 2014 also experienced similar difficulties. Until FGDC effectively oversees the search capabilities on the Geospatial Platform, and improves it as warranted, agencies may continue to experience challenges in finding relevant datasets. In addition, until all federal sources of geospatial data are easily discoverable from one centralized location, the potential for uncoordinated and duplicative data acquisitions remains.

  Although FGDC has recognized the need for performance measures for the Geospatial Platform since 2011—such as numbers of downloads or site visits, percentage of customers who are repeat customers, number of downloads or website hits by user group (e.g., federal, state, or local government, private sector, academic, or citizen users), ease-of-use, and customer satisfaction survey scores—these measures have not yet been implemented. The FGDC Deputy Executive Director also noted that data for determining performance measures will be collected when they are able to implement their new support contact for the Geospatial Platform, which was finalized in October 2014. However, FGDC has not yet established a schedule for when performance measures will be implemented because it is only now starting to work with the support contractor. Until it has a schedule for establishing performance measures, it is difficult to oversee progress on this important step. Moreover, until performance measures are in place, it will be difficult for FGDC to improve the operations of the Geospatial Platform.

- **Partnerships.** FGDC has initiated plans and activities for coordinating with state governments in the collection, production, sharing, and use of geospatial data and the implementation of the NSDI. These efforts include defining desired outcomes for collaboration, clarifying roles and responsibilities for federal agencies, and establishing methods for operating across organizational boundaries.

  However, officials with six of the seven state or state-related entities we met with expressed dissatisfaction with FGDC's efforts. For example, NSGIC officials, who represent the interests of all 50 states, stated that they were frustrated that, although at one time states were allowed to participate in FGDC Steering Committee discussions of geospatial policy and plans that affect them directly, they no longer have a meaningful role because of FGDC's reliance on NGAC. NSGIC officials, officials from the Coalition of Geospatial Organizations, and an official from one of the five states are further concerned by this development because they perceive that the NGAC does not have any real influence over FGDC actions, and that NGAC's advice and recommendations are often ignored. Further, officials from NSGIC, Coalition of Geospatial Organizations, and two of the states in our sample criticized FGDC for operating from a "federal" perspective as opposed to a "national" perspective.

  FGDC officials including the Executive Director, Deputy Executive Director, and Senior Policy Advisor are aware of concerns with their coordination with state governments. The FGDC Deputy Executive Director noted that FGDC plans to improve coordination with state and local officials and to collect data for performance measures on this coordination now that they have a new support contract in place. FGDC officials stated that the support contract's integrated work plan is expected to be issued in early 2015, and is to include tasks and staffing for communication and outreach efforts that they expect to result in better partnerships with state and local governments.

FGDC officials also provided their insights into other state concerns. Specifically, they noted that while the states and FGDC worked closely on technical and foundational issues for the NSDI, the states' direct involvement in FGDC's decision-making processes had to end once the focus shifted to federal policy and budgetary issues due to the sensitive nature of proposed budgets and contracts. The FGDC official noted, however, that the states' interests are still being represented on the committee that advises FGDC. In addition, FGDC officials stated that they made a concerted effort to make the strategic plan encompass a national vision by involving non-federal stakeholders throughout the process and that the plan was unanimously endorsed by the NGAC, including the six NGAC members who represent the interests of the state governments.

Federal Departments and Agencies Have Made Progress on NSDI Initiatives, but Shortfalls Remain and States Have Concerns about Changes in, Restrictions on, and Limitations of Coordination

Selected federal departments and agencies have made progress in complying with OMB guidance to facilitate the creation of an NSDI, including establishing policies for making their data available on the Geospatial Platform, procedures for checking the platform prior to expending funds to collect or produce new geospatial data, and strategic planning mechanisms. However, shortfalls remain in each of these initiatives. Further, while selected state officials voiced praise for some of their partnerships with federal agencies, they also have concerns about changes to coordination programs and restrictions on participating in them. Details on the initiatives follow:

- **Policy on geospatial metadata.** While the federal departments in our prior review (Commerce, Interior, and Transportation) have developed policies that require their agencies to make their geospatial metadata available on the Geospatial Platform, the two additional departments in this review (DHS and USDA) do not have such policies.

In addition, few agencies have made their metadata available on the platform. Of the 32 departments and agencies that make up the FGDC steering committee, 8 departments have made at least a portion of their metadata available on the platform. Further, FGDC identified 188 NGDA datasets as most critical, but as of December 2014, 111 datasets (59 percent) were registered on the platform. Of the six agencies we reviewed that are required to make their data available, five agencies had metadata on the platform and one had not registered any NGDA datasets on the Geospatial Platform.[51] Specifically, Census, FEMA, FSA, NOAA, and USGS had registered NGDA datasets on the platform, but BLM had not—even though the agency is responsible for 4 NGDA datasets.

The agencies provided several reasons for these shortfalls. DHS officials reported that they do not have policies or procedures for putting their metadata on the Geospatial Platform or accessing it before expending funds on geospatial data because they are currently in the process of revising and updating all of the department's geospatial documents and directives. Similarly, USDA officials acknowledged that they do not currently have such policies or procedures, but noted they are in the process of revising their existing guidance on making their metadata publically available and plan to include a reference to posting this information on the Geospatial Platform. Neither agency provided a date for the planned completion of these policies. Until the agencies establish dates for completing their policies, it will be more difficult to oversee and ensure progress is being made.

Officials at BLM stated that they have been trying to publish their metadata on the Geospatial Platform for months, but have been experiencing technical problems in doing so. Other agencies, including the General Services Administration, USGS, and NOAA, reported experiencing similar technical problems. FGDC officials stated that they are working with the Geospatial Platform technical team to address these issues, and have created a metadata working group so that FGDC and affected departments can work together on these problems. FGDC's original goal was to have these problems resolved by December 31, 2014. As of January 2015, FGDC officials stated that they have identified the root causes of the technical issues and have implemented a solution. However, BLM's metadata were still not registered on the Geospatial Platform as of February 2015.

Until all departments have policies requiring them to make metadata available on the clearinghouse and are doing so, the federal government runs the risk of not managing its geospatial data assets in an effective manner. Specifically, if key datasets are not available on the platform, other agencies may not know when they are pursuing potentially duplicative investments. Such situations would result in the inefficient use of already limited resources.

- **Procedures for accessing the clearinghouse.** The federal departments in our prior review (Commerce, Interior, and Transportation) have issued internal procedures to ensure that their agencies access the Geospatial Platform's Marketplace before they expend funds to collect or produce new geospatial data to determine (1) whether the information has already been collected by others or (2) whether cooperative efforts to obtain the data are possible. However, the two additional departments in this review (DHS and USDA) do not have such procedures.

  In addition, while Interior, Commerce, and Transportation have begun to implement these procedures, neither of the other agencies in our current review (DHS and USDA)—nor any of the federal agencies government-wide that are subject to the requirement to do so—are using the Geospatial Platform's Marketplace. Specifically, as of September 2014, three agencies government-wide have registered planned datasets to it: USGS and U.S. Fish and Wildlife Service (Interior), and NOAA (Commerce). While this is an improvement from 2012, when no agencies were using the Marketplace, this is not meeting the intent of the NSDI guidance.

  Agency officials from DHS explained that they do not have procedures for using the Marketplace because it is still too immature. USDA officials stated that are not using the Marketplace because they rely on existing internal procedures and professional networks to identify opportunities for collaboration; however, they are in the process of revising their existing guidance to include a reference to using the Marketplace feature. Using existing professional networks and planning to add a reference to the Marketplace feature fall short of OMB's guidance to ensure agencies access the Marketplace before expending funds to avoid duplicative investments and to enhance coordination.
- Officials with BLM, DHS, USGS, and the National Digital Orthoimagery Program also reported that they find existing personal

and professional networks to be more effective in coordinating geospatial acquisitions than the current Marketplace technology. The NSDI Strategic Plan includes tasks for tracking the use of the Marketplace and documenting associated cost savings and avoidance that are targeted for completion in fiscal years 2015 and 2016.

We acknowledge that the Geospatial Platform's maturity is evolving. However, until agencies use the Geospatial Platform Marketplace to identify planned geospatial data acquisitions, they will likely miss opportunities to reuse or cooperatively acquire geospatial data and increase the likelihood of acquiring duplicative geospatial data.

- **Strategy.** One of the three federal departments in our prior review has prepared and begun implementing a strategy for advancing geographic information and related geospatial activities appropriate to its mission. Specifically, Interior published a plan in March 2014 that addresses the NSDI and includes outcome-oriented goals and objectives and descriptions of how the actions are to be achieved. Progress in implementing the strategy is being monitored by the department's geospatial advisory committee.

  In contrast, the two other departments in our prior review, and the two new departments in this review, have made limited progress in developing such strategies. Specifically, Commerce published a strategic plan in March 2014; however, the strategy does not include descriptions of how its objectives will be achieved, or how progress will be measured. As of January 2015, Commerce officials stated that they plan to develop and approve the implementation plan containing those details by the end of the second quarter of fiscal year 2015. Officials from Transportation, DHS, and USDA stated that they are working on strategies that they anticipate publishing during fiscal year 2015.
- **Partnerships.** Agencies in the federal departments in our review coordinate with states in a number of partnerships and programs, some of which are considered very successful by the states. Programs considered successful by representatives from three of the five states in our sample, as well as NSGIC, include the National Telecommunications and Information Administration's State Broadband Initiative[52] and Transportation for the Nation.[53] Both of these programs were lauded for their ability to work with local governments to build nationwide datasets. Officials from three of the states also commended the NOAA Coastal Services Center,[54] and

officials from four states commended the USGS National Digital Elevation Program,[55] and the U.S. Fish and Wildlife Service's National Wetlands Inventory[56] for their focus on providing products and services that meet states' needs.

However, each of the five states as well as NSGIC expressed concerns about several federal coordination efforts. Specific concerns include (1) changes to USGS programs that have negatively impacted the states, (2) legal and administrative restrictions on sharing federal data, and (3) datasets not being built from the local level up. These concerns are discussed below:

- All five of the state officials and one of the two state organizations we reviewed expressed concerns about changes to partnerships that had been considered successful. One example is the USGS Geospatial Liaison program. This program had established liaisons to engage and support state, local, tribal, and regional partners in improving the timeliness, quality, and accessibility of geospatial data for the community and the NSDI. However, USGS is no longer replacing liaisons that leave the agency or retire, and is reducing travel funding. Further, due to an agency shift in priorities, USGS liaisons are now focusing their efforts on elevation and hydrography, instead of assisting states in coordinating in other areas such as aerial photography.

  Similarly, USGS is ending its long-time support of the Department of Defense's National Geospatial-Intelligence Agency's Urban Area Imagery Program after 2014, again due to its changing focus away from orthoimagery and toward elevation data.[57] The Urban Area Imagery Program was originally created after the terrorist attack on the United States on September 11, 2001, to collect imagery and other geospatial data layers over densely populated urban areas for homeland security and emergency operation purposes. With USGS acting as primary liaison to the states, the program has since expanded beyond its federal scope to become one of the primary mechanisms available to state and local governments for partnering to collect high-resolution imagery over additional urban and non-urban areas as well.
- Each of the five state officials and one of the two state organizations reported that legal and administrative restrictions prevent states from getting the federal data that they need. For example, state officials reported that state and local government agencies work closely with Census to provide updated address data, but that due to legal

restrictions on what Census can share, they do not obtain reciprocal address updates from Census.[58] In addition, state officials reported being hampered in participating in the partnerships to acquire imagery through FSA's National Agriculture Imagery Program.[59] According to state officials, this is because FSA requires payment for the data upfront, while some states' procurement rules require delivery of the product prior to payment. According to FSA, upfront payments allow the agency to make the data available to the states pursuant to federal law.

- All five state officials reported that departments have not always taken advantage of opportunities to integrate local preferences into their data collection efforts as suggested by OMB's guidance.[60] Instead, state officials said that agencies tend to structure their programs to focus on collecting data that support their federal missions. An example of this is the National Agriculture Imagery Program. Some states believe the program would be much more effective and efficient if imagery were collected at the higher resolution preferred by local governments; the images could then be re-sampled to meet the lower resolution requirements of the federal government. Another example where states report that national data would be better collected at the local level, and then rolled up, is address data (which is discussed later in this report).

In response to the state concerns regarding changes to federal programs' coordination with states, officials from USGS noted that funding constraints have forced the changes to the Geospatial Liaison Program and the Urban Area Imagery Program. USGS officials stated while the Urban Area Imagery Program will still be able to use USGS's Geospatial Product and Service Contracts to contract for the data collection, users will have to find other arrangements for the quality assurance and data management activities that USGS can no longer provide. Officials with FSA's National Agriculture Imagery Program noted that state and local governments can still purchase the imagery, or enter into their own contracts for specific imagery products. FSA has a website to facilitate these actions. Given the current budgetary environment, it will be important for agencies to identify and use tools and cooperative agreements that are not heavily reliant on funding.

OMB Has Not Prioritized or Overseen NSDI Efforts

Part of the reason that FGDC and federal agencies are falling short on their responsibilities for the NSDI is that OMB has not made it a priority to ensure that these responsibilities are implemented. While OMB has oversight responsibilities for federal IT systems and acquisition activities—including those that deal with geospatial data—OMB has not made it a priority to oversee agency actions in regard to their geospatial responsibilities. Specifically, although OMB guidance requires agencies to report annually on their strategic planning, agencies have not reported to OMB. Moreover, OMB has not requested agencies provide updates on their progress on NSDI initiatives or provided oversight of agencies' progress. Similar to our prior report, OMB e-government staff told us that this initiative is not a priority for them. Further, while FGDC, with OMB serving as Vice Chair, is required to serve as the lead federal executive body charged with directing and facilitating national implementation of the NSDI, committee officials stated that they are challenged in overseeing the tasks in the strategic plan. Officials stated that this is because they do not have authority over the other federal departments that are responsible for the day-to-day implementation of the NSDI.

As a result, it is now 20 years since the guidance to establish the NSDI was issued and the federal government has not yet fully implemented the NSDI as a mechanism to facilitate the effective and efficient collection, sharing, and dissemination of geospatial data. Unless actions are taken to identify and address federal agency shortcomings, the full realization of the NSDI across all levels of government and the private sector will take even longer, during which opportunities for increased collaboration, reduced duplication, and potential cost savings will be lost.

FEDERAL AGENCIES AND STATES INVEST IN POTENTIALLY DUPLICATIVE GEOSPATIAL DATA; INITIATIVES TO REDUCE DUPLICATION FACE KEY CHALLENGES

Federal agencies and states invest in potentially duplicative geospatial data. While there are initiatives in place to reduce the possibility of duplication, these initiatives face several challenges. OMB guidance encourages the avoidance of duplicative geospatial efforts and investments. In its most recent version of Circular A-16, OMB states that implementation of this guidance is essential to help federal agencies eliminate duplication, avoid redundant expenditures, and improve the efficiency and effectiveness of the sharing and dissemination of geospatial data. Nevertheless, several examples of duplication and the potential for duplication exist in the two types of geospatial data we assessed. Duplication of geospatial data can waste resources; create fragmented and conflicting datasets (in cases where different collection efforts gather outdated or poorly geocoded data); and in some cases where having the correct address data is critical, such as in providing emergency services, risk lives.

Duplicative Address Data Exist within State and Federal Governments; Multiple Initiatives to Reduce Duplication of Address Data Face Challenges

Address data are collected and purchased multiple times by federal, state, and local entities, resulting in duplication of effort and resources. According to NGAC officials, addresses are typically generated at the local level—in a city, town, or county. When state and federal government agencies use address data to support their missions, some go to the local government for the updated address data while other agencies collect their own address data or purchase them from vendors. For example, a developer building a new neighborhood will seek addresses from the local government. Then, the local addressing officials provide updated address data to the USPS, which, in turn, provides a subset of these data to Census and sells address matching services to private vendors.[61] Vendors can then repackage the address data and sell or license them to other organizations, including government agencies. For instance, FEMA licenses address data and geocoding tools from a private vendor, which allows them to geocode the locations of applications for emergency assistance. In addition to using USPS address data, according to agency officials, Census also collects its own supplementary address data, aggregates them from state or local governments and, in some cases, leases or purchases address data from private firms to assist with address research. USPS also reports that it leases address data from vendors to discover new addresses.

In addition to multiple federal and state data collection efforts, address datasets are also sometimes duplicated by agencies within a single state. For example, Washington state's Chief Information Officer estimated that approximately 15 or 16 state agencies have purchased or licensed their own address datasets and that the state would save $1.3 million by eliminating that duplication. That state's GIS Coordinator and Program Manager stated that she believed the potential cost savings to be much higher. See figure 17 for an illustration of duplication in efforts to collect address data.

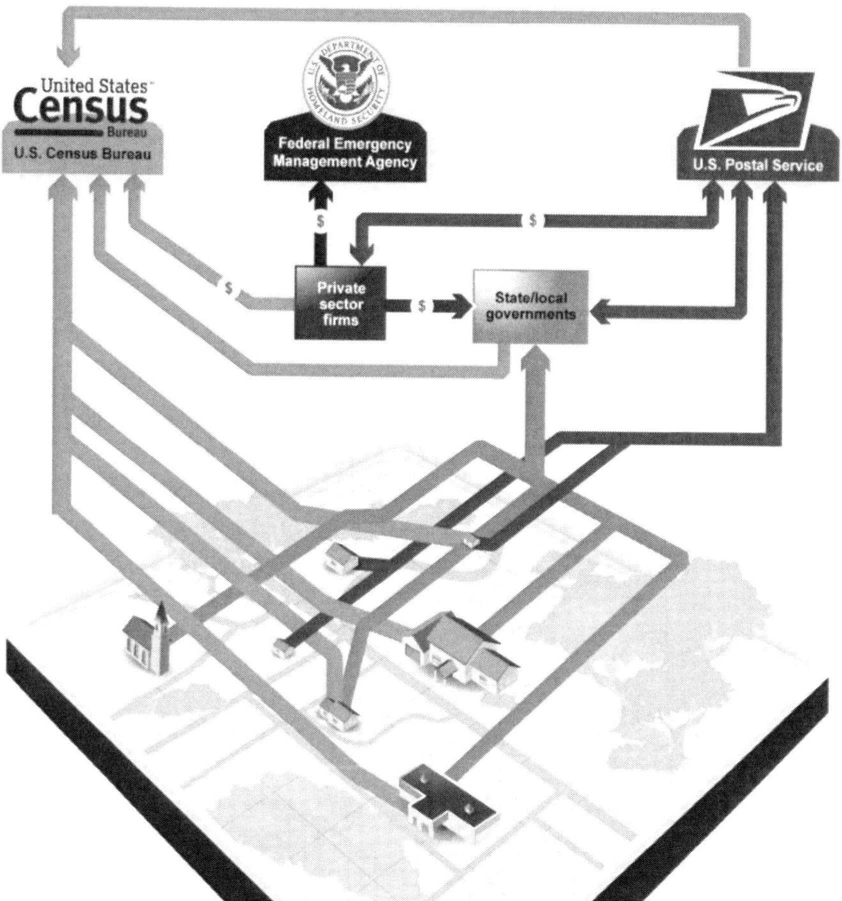

Source: GAO analysis of agency documentation, responses, and interviews. | GAO-15-193.

Note: This is intended to be a simplified visualization and does not capture all address collection efforts done by all federal agencies or reflect Census's current plan to perform reengineered address canvassing in the 2020 Decennial Census.

Figure 17. A Simplified Visualization of Duplicative Efforts to Collect Address Data.

Duplicative or similar datasets can also be inconsistent or inaccurate. Because there are differences in when data were collected, how the data were geocoded, and how often the data are updated, it is not unusual to have different address datasets with inconsistent data. Using inaccurate address data could cause inconvenience, extra work, or—in a worst case scenario—risk lives.

For example, for Census, mailing forms to an inaccurate address could cause the agency to send caseworkers out to visit the location to correct the address. However, for local emergency services entities, having an inaccurate address could cause a delayed response and put lives at risk.

However, it is important to note that many address data collection efforts are not duplicative. In some cases, datasets that appear to be duplicative have customized features that make them distinct. That is, one address file may provide data that another dataset will not, such as geo-locations or parcel data. For example, one of FEMA's datasets includes parcel data, while the other does not. In addition, the intended use of the dataset may differentiate the datasets.

For example, Census maintains three address datasets, but each has its own intended use, such as maintaining a comprehensive list of addresses for all the governments within the United States to assist in the census of governments or a file that has an address for each living quarter in order to conduct the census.

Even so, it is evident that there is duplication of effort, resources, and spending between various levels of government within the address data category. For example, almost any effort to collect or purchase address data would duplicate Census's past and future efforts to collect, update, and verify its database of millions of addresses.

In addition, entities often maintain several purchased address datasets from different vendors. For instance, USPS has five contracts dealing with address data. USPS has recognized this and plans to consolidate the contracts to a single contract in the future. The agencies and states that we reviewed maintained 29 address data files that cost up to $1.4 billion over 10 years. One state reported approximately 13 different address files within its state government alone. Table 7 provides the address datasets created or used by each of our selected agencies and states.

Table 7. Address Data Created or Used by Selected States and Federal Agencies

Organization	Dataset	Use	Coverage area	Source of data	Estimated cost
Census	Master Address File	Conducting the decennial census, as well as other census products	Nationwide	Collected and aggregated	$1.4 billion over 10 years for the 2010 Census[a]
	Geographic program participants database	Providing contact information for participants in census programs	Approximately 39,000 state and local governments' contact information	Collected	Unable to quantify
	Governments Master Address File	Providing contact information on governmental participants	Nationwide—state/local governments	Collected	Unable to quantify
FEMA	Licensed address data within software	Geocoding locations	Unknown	Purchased	$12,589
	Licensed address data within software	Standardizing and correcting addresses	Nationwide	Purchased	$710,018
USPS	Address Management System	Delivering mail to every address in the country	Nationwide	Aggregated and purchased	$36 million/yr for operations and maintenance
	Licensed address data within software	Providing a locator function on USPS.com to find post offices	Unknown	Purchased	$39,383
	Licensed address data within software	Creating maps that mail carriers follow on route	Nationwide	Purchased	$4.3 million
	Licensed address data within software	Managing mail carrier routes	Unknown	Purchased	$358,630
	Licensed address data within software	Preparing for the effects of disasters on mail delivery	Unknown	Purchased	$65,770
	Licensed address data within software	Ensuring package delivery and manage carrier performance	Nationwide	Purchased	$1.8 million

Maryland	Address point locations	Assisting all state agencies in completing their missions, including health, human resources, education, and public safety	Statewide	Aggregated	$75,000/yr
Montana	Statewide address database	Assisting state agencies in completing their missions, including natural resource planning, disaster recovery, and delivering government services	Statewide	Aggregated	$80,000/yr
Ohio	Location Based Response System County Addresses Database	Assisting state agencies in completing their missions, including emergency response, tax determinations, geocoding, and crash analysis	Statewide—80 of 88 counties	Aggregated	Approximately $16 million to develop $400,000 in FY13 and FY14
Ohio	Statewide Master Address File	Assisting state agencies in completing their missions, including emergency response, tax determinations, geocoding, and crash analysis	Statewide	Aggregated	$0
South Carolina	State Address Point Program	Assisting state agencies in completing their missions, including crime mapping, homeland security, research and statistics, and voter validation	Statewide—38 of 46 counties	Aggregated	$6.2 million to develop
Washington[b]	13 different address files	Assisting state agencies in completing their missions, including tracking cancer clusters, determining location of public health services, and issuing permits	Various	Aggregated and purchased	Unable to quantify

Source: GAO analysis of agency documentation, responses, and interviews. | GAO-15-193

Notes:

[a]U.S. Census Bureau, *High-Quality Maps and Accurate Addresses Are Needed to Achieve Census 2020 Cost-Saving Goals*, OIG-12-024-1 (Washington, D.C.: May 10, 2012).

[b]According to state officials, Washington does not currently have a single master address file. Many of its departments maintain their own address data. Washington is in the process of creating a master address file for the state.

Acronyms and abbreviations: Census = U.S. Census Bureau; FEMA = Federal Emergency Management Agency; USPS = United States Postal Service.

Efforts Are Under Way to Reduce Duplication and Increase the Sharing of Address Data

Several efforts are under way to reduce duplication and increase address sharing across governments, including an initiative to create a national address database; a Census initiative to collect address data from state and local governments; and an initiative to modernize the 9-1-1 emergency calling system, called Next Generation 9-1-1.

- **National Address Database Initiative.** At FGDC's request, in December 2012, NGAC published a whitepaper on the need for developing a national address database.[62] Acknowledging that address data are an essential requirement for a wide variety of governmental functions (including emergency management, administration, research, publications, mapping, routing, and navigation), NGAC reported that there is a critical need for a national address database as a single repository for storing, aggregating, and sharing essential address information.

  NGAC's vision for a national address database was a continuously updated, nationwide, publicly available address database that meets the needs of governments at all levels where address data are developed locally, with local and state custodians acting as regional integrators. The report provided several potential alternatives for the development and the management of a database, including housing the database within one federal agency or having several stakeholder agencies share responsibility for the database.

  NGAC reported that the benefits of a national address database would be improved ability to respond to emergency and public safety incidents, improved government services, less fragmented address systems, and cost savings. Further, in December 2014, NGAC identified seven user scenarios to demonstrate the value of a national address database, including federal, state, and local government uses for activities such as emergency response and voter registration. While NGAC did not define the total potential cost savings, the amount could be quite large. For instance, if the address data were improved, Census could move to more targeted address canvassing and potentially save $196 million on the 2020 census. Further, DHS estimated a cost avoidance of $2 million per year for address data access. This did not account for the benefit of time savings for incident response such as survivor identity validation and eligibility, individual housing assistance, damage assessments, and public assistance. Given the number of federal, state, and local agencies that collect and maintain address data on an ongoing basis, it is clear that a coordinated and comprehensive national address database could potentially lead to significant savings. Going forward, NGAC recommended that (1) FGDC add addresses as a new data theme and identify a theme lead agency and (2) FGDC agencies coordinate and collaborate to develop a business requirements plan for a national address database during fiscal year 2015.

There are several potential challenges to moving forward on a national address database. First, according to NGAC officials there are questions regarding where the national database should reside and which agency should lead the effort. Further, given statutory limits affecting Census and USPS, NGAC officials expressed concern that the agencies' involvement could limit the access to the information.[63] Additionally, privacy professionals have expressed a concern regarding whether a building's physical address is personally identifiable information, and therefore private.[64] NGAC has recognized these concerns and proposed that the national address database should only contain addresses and not identities of individuals, owners, or occupants. Officials noted that buildings have no right to privacy.

- **Geographic Support System Initiative.** In fiscal year 2011, Census initiated its Geographic Support System, a $674 million, 10-year project. As part of this project, Census's Geography Division is working on an initiative with USPS; other federal agencies; and state, local, and tribal governments to allow them to regularly share and continuously update their address lists and road data with Census. The Chief of the Geography Division estimated that, as of August 2014, they had collected data for about one-third of the population. Census anticipates collecting the data throughout the decade leading up to the 2020 Decennial Census. Census has estimated that this initiative may yield a reduction in the cost of address canvassing throughout the nation on the order of $1 billion.

  However, we recently reported that Census's lack of milestones, deadlines, and measurable goals with regard to this effort make it difficult to know whether the bureau is on track to meet its goals.[65] We recommended the agency develop a detailed plan for its data sources for the 2020 Decennial Census that includes measurable goals, track performance against these goals, and set a timeline. The agency generally agreed with our recommendations.

- **Next Generation 9-1-1.** We recently reported that public safety entities are in the process of implementing the next generation of 9-1-1 services to, among other things, improve their capabilities to communicate with callers, increase resiliency of their 9-1-1 operations, and enhance information sharing among first responders.[66] Next Generation 9-1-1 is expected to use Internetprotocol-based, broadband technology that is capable of carrying voice plus large amounts of varying types of data, such as instant messaging, wireline calls, voice over Internet protocol calls, photographs, and live video feeds from an emergency scene.

  The original 9-1-1 system was designed to carry only the caller's telephone number with the call, and the associated fixed address was obtained from an established database. However, the mobility of wireless callers makes fixed street addresses unreliable as location indicators. Next Generation 9-1-1 is to map the caller's geospatial coordinates and route the call to the appropriate Public Safety Answering Points—the locations answering 9-1-1 phone calls. Some local and state governments are using this opportunity to develop centralized address databases that share information across departments and with the local 9-1-1 authority. While this initiative holds promise, it is not clear when it will be completed.

Agencies and States Face Challenges in Preventing Duplication in Address Data

Agencies and states report facing two key challenges in preventing duplication in address data collection efforts. They include limited sharing of address data due to statutory restrictions and a lack of federal sponsorship.

- **Statutory restrictions.** While Census has a massive address database that must be updated with new address data on an ongoing basis, agency officials stated that the agency is limited in sharing the address data it collects due to statutory restrictions under Title 13.

  Specifically, to protect privacy, Section 9 of Title 13 of the U.S. Code prohibits Census from disclosing data it collects about individuals and establishments in a manner that would identify those individuals and businesses. Consequently, Census can collect address data from multiple sources, but it believes it cannot share those data with other federal or state government agencies. For example, the Census Address List Improvement Act of 1994 authorized USPS to share its address data with Census; however, Census does not share its address data with USPS. Similarly, Section 412 of Title 39 of the U.S. Code prohibits USPS from disclosing names or addresses (past or present) of postal patrons or other persons.

This lack of sharing leads to duplication of effort and inefficiency. For example, in 2009, Commerce's National Telecommunications and Information Administration began its state broadband initiative to allow users to view broadband availability across the United States. The agency awarded grants to each state or the states' designee to assist in gathering data on the availability, speed, and location of broadband services. In order to gather those data by address, 17 states used $13 million in agency-provided grants to create, update, or complete statewide address databases. Had the address data been already available, those funds could have been directed elsewhere. A reexamination of the restrictions on address data could allow for greater coordination and reduce duplicative efforts.

- **Federal sponsorship.** Another challenge agencies face is that there is no federal sponsorship at the national level for address data. While both NSGIC and NGAC have an address subcommittee, according to FGDC officials, FGDC does not have a similar subcommittee to address these particular issues because it does not currently recognize address data as its own data theme. FGDC's Executive Director stated that this was because there was no specific agency champion to serve as theme lead. Both NSGIC and the Urban and Regional Information Systems Association[67] have petitioned Interior and FGDC to include address data as its own data theme noting that its addition would help reduce duplication.

In the absence of a single, authoritative, and publicly available address database, multiple agencies from all levels of government are building, maintaining, and/or paying for multiple address databases. The result is inconsistent data, redundant business processes, and wasted taxpayer dollars. In addition, since addresses are the most commonly used way to communicate the location of an emergency, the results of inaccurate or incomplete address data could affect emergency response and public safety. Until there is increased focus on building a national address database and providing federal sponsorship for that effort, there will continue to be duplicative address datasets developed at every level of government.

Current Aerial Photography Datasets Appear to Have Limited Duplication, but Opportunity Exists for More Coordination

Selected agencies and states have multiple aerial photography datasets, which appear to have limited duplication. Specifically, selected agencies and states reported having 12 aerial photography datasets. The datasets differ mostly due to specific agency and state business needs requiring varying resolutions and coverage area. Two organizations may both need imagery data, but for different purposes, using different definitions and different levels of precision or accuracy, such as resolution. An example of this is the National Agriculture Imagery Program and states' own collection of imagery. While the National Agriculture Imagery Program does have imagery of the states, such as South Carolina, it is not always helpful to the states to use the program's imagery. According to South Carolina officials, when trees are thick with foliage, the program's leaf-on imagery does not allow mapping of certain features, such as streets, streams, and houses. Additionally, South Carolina (like many states) requires a higher resolution for imagery so that certain features can be mapped closer to the ground (e.g., road features and houses). However, according to state officials, some states, like Washington and Montana, depend on the program to provide the state's only statewide aerial photography. Table 8 provides a summary of aerial imagery collected and acquired by selected state and federal agencies.

Programs That Support the Coordination of Aerial Photography Face Challenges That Could Result in Duplication

To coordinate the acquisition and collection of imagery, several imagery programs have been established and evolved to include all levels of government, while making low costs and public access a priority. These programs include the National Digital Orthoimagery Program, the USGS Geospatial Liaison Network, the National Agriculture Imagery Program, and the 133 Major Urban Areas Program. These intergovernmental partnerships to acquire imagery have clear advantages: lower costs, reduced duplication of effort, greater standardization, and more data available to users.

Table 8. Aerial Imagery Collected or Acquired by States and Federal Agencies

Organization	Dataset	Collection cycle	Leaf-on/leaf-off	Resolution	Coverage area	Source of data	Estimated cost for FY13 and FY14
BLM	Oregon/Washington Cyclic Aerial Photography	5 years	Leaf-on	24 inch	Oregon	Purchased	$337,698
	Multiple individual aerial photography projects	Varies	Varies	Varies	Varies	Purchased	$1.3 million[a]
FSA	National Agriculture Imagery Program	2-2.5 years	Leaf-on	1 meter	States in the continental U.S.	Purchased	$9.6 million[a] $10.1 million[b]
NOAA	Coastal Aerial Imagery	5-15 years based on susceptibility to change	Leaf-off is preferred	35 centimeter	Coastal regions	Collected, leveraged, and purchased	Not provided
NOAA	Emergency Response Imagery	Event-driven	Event-driven	25 centimeter	Event-driven	Collected, aggregated, and purchased	Not provided
National Geospatial-Intelligence Agency (with USGS)[d]	High Resolution Orthoimagery/133 Urban Areas	2 years	Leaf-off	1 foot	Urban areas nationwide	Purchased and collected	$5,091,855[a] $4,035,856[b] (just USGS costs)

Maryland	High Resolution Aerial Imagery	3 years	Leaf-off	6 and 3 inch	Statewide	Purchased	$815,939[a] $810,062[b]
Ohio	High Resolution 4-band Aerial Imagery	3 years or when funds available	Leaf-off	1 foot, with 3 or 6 inches for various counties	Statewide	Purchased	$452,000[a] $418,000[b]
	Oblique Imagery	As needed by locals	Leaf-off	3-8 inch, depending on rural or urban setting	Statewide	Purchased	$0[c]
South Carolina	High Resolution Orthoimagery	10 years	Leaf-off	6 or 12 inch	County-by-county basis	Leveraged	$340,000 (local imagery) $210,000 (federal imagery)
	Oblique Imagery	One-time or every other year	Unknown	3-12 inches	Selected counties	Purchased	Not provided
	Ortho Quarter Quadrangles	4-6 years	Leaf-off	1 meter	Statewide	Leveraged	$0[c]

Source: GAO analysis of agency documentation, responses, and interviews. | GAO-15-193

Note: Washington and Montana currently do not have statewide aerial photography programs and therefore they rely on the National Agriculture Imagery Program.

[a]Expenditure is from fiscal year 2013. [b]Expenditure is from fiscal year 2014.

[c]No expenditure in fiscal years 2013 and 2014.

Acronyms and abbreviations: BLM = Bureau of Land Management; FSA = Farm Service Agency; NOAA = National Oceanic and Atmospheric Administration; USGS = U.S. Geological Survey.

However, these programs face challenges, including changing priorities and limited resources for coordination efforts that could reduce collaboration and therefore potentially result in duplication. A summary of each program and its challenges follows.

- **The National Digital Orthoimagery Program.** The National Digital Orthoimagery Program is a consortium of agencies that acts as the FGDC subcommittee for imagery. While its members are principally federal—including the USGS, FSA, Natural Resources Conservation Service, U.S. Forest Service, BLM, FEMA, Census, and NOAA—state governments are represented by NSGIC. The program's goal is to provide national orthoimagery coverage by combining funding resources and creating partnerships to coordinate requirements and costs with federal, state, other government agencies, and the private sector. According to program officials, their outreach efforts ensure that state and local governments are always aware of opportunities to partner with federal government agencies.
  However, none of the states included in our review reported having any direct or recent interactions with the program. According to National Digital Orthoimagery Program officials, this could be attributable to the fact that, due to budget constraints, it is no longer able to conduct its bi-annual meetings at locations where opportunities to build relationships with state and local officials could occur.
- **USGS Geospatial Liaison Network.** As previously stated, USGS Geospatial Liaisons engage and support state, local, tribal, and regional partners in improving geospatial data. Specifically, USGS Geospatial Liaisons develop partnerships and agreements that improve the data holdings and the products and services of The National Map, including imagery. For example, in 2012, USGS Geospatial Liaisons helped to facilitate 52 agreements for leaf-off high-resolution imagery in 35 states. However, a declining budget and change of priorities have caused significant changes to the liaison program. Major changes to the program include a reduction in the number of liaisons and a change in priorities to focus primarily on the elevation and hydrography themes, rather than assisting with any other geospatial issues, including aerial photography. By focusing on elevation-related issues, states can no longer rely on their liaison to coordinate shared aerial photography partnerships.

- **National Agriculture Imagery Program.** FSA funds and administers the National Agriculture Imagery Program, which acquires 1-meter aerial imagery during the agricultural growing seasons in the coterminous United States in a 2-year cycle. Other agencies share about 36 percent of the costs.[68] The resulting digital orthophotography is available to government agencies and the public to download at no charge. FSA allows federal, state, and local agencies to partner for more detailed orthophotography collections, known as "buy-ups," subject to FSA approval and at additional costs.
  Officials from four of the states in our review report that they rely on the 1-meter imagery, and NSGIC officials consider the program a great success. However, it is challenging for states and some agencies to participate in the buy-ups that would provide for the more detailed orthophotography. Specifically, according to agency officials, the buy up program requires upfront payment, which is not always possible in the budget environment of states. Officials with three of the five states in our sample noted that FSA's requirement that contracting partners pay in advance for specialized data collections precludes them from participating in the buy-ups, because their states' procurement rules prohibit payment in advance of receipt of a product.
- **133 Major Urban Areas Program.** Coordinated through USGS and the National Geospatial-Intelligence Agency, the 133 Major Urban Areas imagery program acquires imagery for 133 urban areas of the United States to meet critical homeland security and emergency services requirements at a more detailed resolution than that of the National Agriculture Imagery Program—1-foot imagery, as opposed to 1-meter imagery. This effort leverages the resources of multiple partners to achieve significant savings through economies of scale. The agency and USGS have also coordinated with state and local agencies to address their special needs when possible, resulting in cost savings. However, in 2014, due to limited resources and changing priorities, USGS ended its support for the program. As a result, the program's future is uncertain.

In an attempt to increase partnership between federal, state, and local governments, in 2004, NSGIC proposed the Imagery for the Nation initiative, which was intended to provide a sustainable and flexible national digital imagery program that would meet the needs of most government agencies by collecting and disseminating standardized multi-resolution products. The program was to combine the National Agriculture Imagery Program and an effort to collect 1-foot, leaf-off imagery for the entire nation in order to leverage the resources of multiple partners (including state and local governments) and achieve significant cost savings for all parties and improve the federal government's imagery collection. In 2007, USDA and USGS conducted a cost-benefit analysis on the program that evaluated several alternative concepts to implement it. The analysis showed that the program could save federal, state, and local governments up to $938 million dollars over 10 years. By 2009, FGDC had acknowledged the need for a national imagery program and initiated a project to define the federal role in the program. Further, in 2010, FGDC, in collaboration with the National Geospatial-Intelligence Agency, issued a request for information to determine the most effective and efficient options for the program.

However, despite the promise of considerable cost savings, the program's progress stalled in 2011. According to the Executive Director of FGDC, the initiative is now dormant due to a lack of funding and a lack of interest in reviving the program. Contrary to this, officials from several states stated that they would use such a program and expressed disappointment that it never got started. A Maryland state official reported that the state has implemented such an imagery program by consolidating the funding received from an Enhanced 9-1-1 trust fund to conduct a statewide aerial photography acquisition that meets the needs of the state's counties. The official noted that this approach has been very efficient and effective in getting the counties the level of data they need, while providing the state with one coordinated dataset, resulting in cost savings to both levels of government.

Given the benefits that a more comprehensive approach to national imagery offers and the current budgetary environment, it is more important than ever for the federal government to create efficiencies and cost savings by taking advantage of intergovernmental partnerships. Without a national imagery program that incorporates the needs—as well as the resources—of all levels of governments, increasingly scarce resources will not be spent efficiently and effectively in obtaining data critical to the missions and operations of our federal, state, and local governments.

CONCLUSION

While the use of geospatial data by the federal and state governments is ubiquitous and federal agencies and states report spending significant funds on these data, the estimated costs to the taxpayer are significantly underreported. In 2012, we recommended that OMB develop a mechanism to identify and report on all geospatial-related investments, and efforts to address that recommendation are under way. In addition, a recently passed law—the Digital Accountability and Transparency Act of 2014—could provide OMB with another opportunity to improve information about federal spending on geospatial data and investments.[69]

Initiated in 1994 to coordinate investments in geospatial data, the NSDI still holds promise; however, more remains to be done to fully implement it and reap its benefits. In fulfilling its responsibilities for the NSDI, FGDC has initiated strategic planning and portfolio management practices and established the Geospatial Platform as a clearinghouse for sharable datasets. However, these efforts have not yet reached a level of maturity necessary for effectively implementing the NSDI. Specifically, the search capability within the clearinghouse does not always identify relevant datasets, FGDC does not yet have a schedule for when it will implement performance measures for the clearinghouse, and state organizations believe that they do not have a meaningful role in the NSDI. In addition, the federal departments and agencies in our review have made only limited progress in complying with their NSDI responsibilities, such as making their data available on the Geospatial Platform, checking the platform prior to expending funds to collect or produce new geospatial data, and developing strategic planning mechanisms. Specifically, Commerce and Interior established policies for identifying their geospatial datasets on the platform, and procedures for accessing the platform prior to investing in new geospatial data, while USDA and DHS did not. Both agencies are revising their existing policies, but have not yet set schedules for implementing them.

Until FGDC is able to provide a mature foundation for the NSDI (including having an effective search engine and a schedule for measuring performance), federal departments and agencies (including USDA and DHS) are fully implementing their responsibilities by establishing a schedule for when key policies will be implemented, and all federal stakeholders are working effectively with states to collect data of national interest, the vision of the NSDI to improve the coordination of geospatial data will not be fully realized, and limited resources will likely not be used effectively.

One reason that FGDC and agencies are lagging in their efforts to establish the NSDI is a lack of oversight. While FGDC provides guidance, it does not have authority to oversee other federal departments in their geospatial responsibilities. OMB, pursuant to its responsibility to ensure the efficient and effective use of information technology, is in a position to oversee agency efforts, but has not made it a priority to provide this oversight. Until OMB provides this much-needed oversight, it is likely that multiple agencies will continue to delay efforts to populate and use the Geospatial Platform. As a result, key opportunities for improving coordination and reducing duplication will be missed.

While OMB guidance states that agencies are to eliminate duplication, avoid redundant expenditures, and improve the efficiency and effectiveness of sharing and dissemination of geospatial data, several examples of duplication and the potential for duplication exist within address data. Specifically, address data are collected multiple times by federal, state, and local entities, resulting in duplication in effort and resources. Several efforts are under way to reduce duplication and increase address sharing across the government, including an initiative to create a national address database. However, agencies and states face challenges in effectively coordinating these efforts, such as limited sharing of address data due to statutory restrictions and the lack of federal sponsorship. Further, selected agencies and states have multiple aerial photography datasets, which appear to be similar but not duplicative, mostly due to efforts to coordinate the acquisition and collection of imagery. However, these programs face challenges, including declining resources and changing priorities, which could reduce coordination and therefore potentially result in duplication. Until there is consistent coordination across the national geospatial data infrastructure, there will continue to be duplicative datasets developed at every level of government, resulting in wasted resources, fragmented and conflicting datasets, and in some cases, risked lives.

MATTER FOR CONGRESSIONAL CONSIDERATION

To increase coordination between various levels of government and reduce duplication of effort, resources, and costs associated with collecting and maintaining accurate address data, Congress should consider assessing the impact of the disclosure restrictions of Section 9 of Title 13 and Section 412 of Title 39 of the U.S. Code in moving toward a national geospatial address database. If warranted, Congress should consider revising those statutes to authorize the limited release of addresses, without any personally identifiable information, specifically for geospatial purposes. Such a change, if deemed appropriate, could potentially result in significant savings across federal, state, and local governments.

RECOMMENDATIONS FOR EXECUTIVE ACTION

To better facilitate the coordination of—and accountability for—the estimated billions of dollars in federal geospatial investments, and to reduce duplication, we are making a total of nine recommendations to OMB, FGDC, and two federal departments. Specifically, to make progress toward an effective national infrastructure and to improve oversight on federal spending on geospatial data and assets, we recommend that the Director of OMB take the following action:

- Improve oversight of progress on the NSDI by requiring federal agencies to report on their efforts to establish and implement policies for identifying geospatial metadata on the Geospatial Platform and their procedures for utilizing the Marketplace feature of the Geospatial Platform before making new investments in geospatial data.

In addition, we recommend that the Secretary of the Interior, as the FGDC Chair, direct the FGDC Steering Committee to take the following four actions:

- Correct the search function of the Geospatial Platform so that it effectively identifies applicable datasets, and oversee the effectiveness of the search function on a regular and ongoing basis.
- Establish a schedule for collecting performance measures for the Geospatial Platform and report on these measures during Steering Committee meetings.
- Create an address data theme with associated subcommittees and working groups to assist in furthering a national address database.
- Direct the National Digital Orthoimagery Program to reassess the feasibility of the "Imagery for the Nation" initiative, with the goal of identifying discrete steps that could be taken to further a national imagery program benefitting governments at all levels.

To help ensure the success of departmental efforts to improve geospatial coordination and reduce duplication, we recommend that the Secretary of Agriculture direct the designated senior agency official for geospatial information to take the following two actions:

- Establish a schedule for developing and implementing a policy that requires the department to make its geospatial metadata available on the Geospatial Platform.
- Develop and implement internal procedures to ensure that it accesses the Geospatial Platform Marketplace before it expends funds to collect or produce new geospatial data to determine (1) whether the information has already been collected by others and (2) whether cooperative efforts to obtain the data are possible.

To help ensure the success of departmental efforts to improve geospatial coordination and reduce duplication, we recommend that the Secretary of Homeland Security direct the designated senior agency official for geospatial information to take the following two actions:

- Establish a schedule for developing and implementing a policy that requires the department to make its geospatial metadata available on the Geospatial Platform.
- Develop and implement internal procedures to ensure that it accesses the Geospatial Platform Marketplace before it expends funds to collect or produce new geospatial data to determine (1) whether the information has already been collected by others and (2) whether cooperative efforts to obtain the data are possible.

AGENCY COMMENTS AND OUR EVALUATION

We requested comments on a draft of this report from the four agencies to which we made recommendations (OMB, Interior, USDA, and DHS), as well as from the three agencies that did not receive recommendations (Commerce, GSA, and USPS). OMB, Interior, USDA, and DHS generally agreed with our recommendations. Commerce agreed with our findings, and provided technical comments. GSA responded that it did not have comments on the draft report. USPS did not agree or disagree with our findings, but provided comments. Each agency's comments, where applicable, are discussed in more detail below:

- In e-mail comments provided by the Chief of Policy, Budget, and Communications of the Office of E-Government and Information Technology on January 16, 2015, OMB generally agreed with our recommendation and stated that it appreciated our careful review of federal geospatial data and associated spending, agency progress in establishing the NSDI, and analysis of potential challenges agencies face in coordinating geospatial data collection efforts.

- In written comments, Interior generally agreed with our recommendations, but noted that there were some key points missing from the report. Specifically, the agency stated that until the term "geospatial data costs" is defined and a consistent method and set of investment assets for reporting are defined and consistently applied, it is conjectural to make general statements of over- or under-reporting of costs. While we agree that "geospatial data costs" needs to be better defined and that a mechanism for collecting the data needs to be consistently applied, we believe it is appropriate to assess existing OMB-required reporting mechanisms to gain insight into what agencies report they are spending on geospatial investments. Further, we believe it is appropriate to make the case that spending on geospatial investments is underreported by identifying programs that clearly gather or use geospatial data that are not captured by OMB's current reporting mechanisms. As previously stated, OMB and FGDC have made plans to improve cost tracking to address a previous GAO recommendation.

  In addition, Interior stated that while federal and non-federal partnerships can be very successful, often the partnerships rely on federal funding. Interior further noted that in the funding environment of the last several years, these types of investments have been predominantly cut, and agencies have had to reduce or eliminate some of these partnerships. While the majority of the partnerships we reviewed were most successful when they had stable funding, in the current budgetary environment, it is important for agencies to identify and use tools and cooperative agreements that are not heavily reliant on funding.

  Further, in technical comments accompanying its letter, Interior generally agreed that congressional assessment of Section 9 of Title 13 of the U.S. Code could potentially result in cost savings across government. In addition, the agency stated that the Census Bureau and the Department of Transportation are collaborating on an analysis of requirements for a national address database, and that the need for and approach to creating an address data theme and its associated subcommittee will be evaluated as part of the analysis and submitted to the steering committee for consideration.

- In written comments, USDA agreed with our findings and stated that the agency intends to move forward with actions to complete the two recommendations. Specifically, the agency plans to include appropriate agency-level geospatial data policy guidance and procedure best practices in its governance documents. In addition, the agency acknowledged the value of reassessing deployment alternatives for "Imagery for the Nation," which it continues to endorse.
- In written comments, DHS concurred with the recommendations and stated that it plans to update existing geospatial management policy and procedures to reflect the recommendations by September 30, 2015. DHS also provided technical comments, which we incorporated as appropriate.
- In e-mail comments provided by a Commerce liaison on January 15, 2015, Commerce concurred with the general findings as they applied to the Department of Commerce. Commerce also provided technical comments which we incorporated as appropriate.
- In written comments, USPS did not agree or disagree with our findings. However, it provided several comments on the report. Specifically, USPS provided comments explaining its address products, and as a result, we clarified the language in our report that describes these products. Further, USPS requested that we expand the matter for congressional consideration so that Congress would also consider assessing the statutory requirements that impact USPS—Section 412 of Title 39. We considered including Title 39 in our matter for congressional consideration and agree that USPS's suggestion has merit. As a result, we modified our matter for congressional consideration to include an assessment of Title 39.

David A. Powner, Director, Information Technology Management Issues

APPENDIX I: OBJECTIVES, SCOPE, AND METHODOLOGY

The objectives of our review were to (1) describe the geospatial data that selected federal agencies and states are using to support their missions, and identify how much is spent on geospatial data; (2) assess progress in establishing the National Spatial Data Infrastructure (NSDI); and (3) determine whether selected federal agencies and states invest in potentially duplicative geospatial data.

To address our objectives, we selected two key data types, seven federal agencies, and five states. We selected the two key data types— addresses and imagery resulting from aerial photography—based on (1) the need to appropriately scope of our review due to the large amounts of geospatial data, (2) the importance of these types of data to the work of the federal government, and (3) the potential for duplication within these data types, as noted in prior federal reports.

We selected agencies based on three factors. Specifically, we selected agencies whose missions were closely tied to the selected data types (addresses and aerial photography). In further selecting from among these agencies, we sought a mix of agencies who participate in the geospatial clearinghouse[1] and those who do not. In addition, we sought to include departments that were not the subject of our prior geospatial report.[2] The federal agencies we selected because of their reliance on address data are the Department of Commerce's (Commerce) Census Bureau (Census), the Department of Homeland Security's (DHS) Federal Emergency Management Agency (FEMA), and the United States Postal Service (USPS). The federal agencies we selected because of their reliance on aerial photography are the U.S. Department of Agriculture's (USDA) Farm Service Agency (FSA), Commerce's National Oceanic and Atmospheric Administration (NOAA), and the Department of the Interior's (Interior) Bureau of Land Management (BLM) and U.S. Geological Survey (USGS). As of November 2013, when we determined the scope of this engagement, four of these agencies (Census, FEMA, NOAA, and USGS) participated in the clearinghouse and the other three did not.[3] Further, Interior and Commerce were included in our prior review, while USDA, DHS, and USPS were not.

We selected the states in our review based on whether they were reported to have both state imagery and address data programs, and whether the state hosted federal imagery programs (so that we could determine whether duplication existed between federal government and states). In addition, we sought states that would provide regional diversity within the United States. The selected states are Maryland, Montana, Ohio, South Carolina, and Washington. To describe the geospatial data that the selected federal agencies and states are using to support their missions, we reviewed documentation regarding federal and state use of geospatial data, including Congressional Research Service reports, the National States Geographic Information

Council's (NSGIC) data inventory, the Federal Geographic Data Committee's (FGDC) list of National Geospatial Data Asset (NGDA) datasets, FGDC's Geospatial Platform and its recent annual reports, and attended a geospatial data users conference. In order to assess the reliability of the FGDC list of datasets, we traced the datasets back to the Geospatial Platform, reviewed FGDC and Office of Management and Budget (OMB) guidance on registering datasets on the platform, and asked agency officials to verify their datasets. In order to assess the reliability of the NSGIC inventory, we reviewed selected states' datasets, traced the datasets back to the Geospatial Platform, and interviewed NSGIC officials. We also interviewed the selected agencies' and states' officials to determine which types of geospatial data they were using and how they were using them. We determined that these data were sufficient for our purposes. To identify how much is spent on geospatial data, we analyzed agency-reported data on IT investments and acquisitions from the IT Dashboard and USASpending.gov. As part of that analysis, we isolated the costs of geospatial data by filtering results by the geospatial categorization (where available). We also requested cost estimates on geospatial data from state and agency officials. To assess the reliability of the cost data, we searched the IT Dashboard and USASpending.gov website for investments that were not categorized as geospatial by using targeted key word searches and comparing those results to those that were categorized as geospatial. We also reviewed the results of previous GAO reports on the IT Dashboard and USASpending.gov that had identified deficiencies in the accuracy and reliability of the reported data.[4]

To assess the reliability of agency and state cost estimates, we reviewed supporting documentation, where available, and interviewed state and agency officials on the sources of their estimates and their data controls. We also had the agencies verify the results. We determined that the cost data from the two websites and from the agencies, while incomplete, were sufficient for our purpose of demonstrating the lack of comprehensive data on the cost of geospatial data. To assess progress in establishing the NSDI, we focused on activities being undertaken government-wide by the FGDC and on activities undertaken within the departments selected for this review. We reviewed FGDC and federal department and agency documentation, such as policies, procedures, strategic plans, implementation plans, and program information. We assessed this information against responsibilities identified in Executive Order 12906,[5] OMB Circular A-16,[6] and OMB M11-03;[7] identified any discrepancies and discussed them with relevant agency officials. We also followed up on activities being taken by federal departments and agencies that had open recommendations from our prior review,[8] which included the

Department of Transportation and the Bureau of Transportation Statistics, in addition to Commerce, NOAA, Interior, and USGS.

In order to determine the progress being made in establishing a clearinghouse, we evaluated the effectiveness of the Geospatial Platform's search function by extracting 1,054 aerial photography and 10,272 address metadata records from the Geospatial Platform. We then randomly selected a generalizable sample of 178 aerial photography and 192 address metadata records. To assess the reliability of the records, we reviewed documentation on the Geospatial Platform's architecture, traced the records back to the platform, and interviewed Geospatial Platform officials. We determined that the data were sufficient for our purpose of assessing the functionality of the Geospatial Platform. To determine if each selected record appeared in search results, we searched for these datasets on the Geospatial Platform using the exact names of the datasets as they appeared in the catalog.

To determine the extent to which federal agencies have made their metadata available on the Geospatial Platform, we conducted searches of the platform, filtered by agencies and reviewed FGDC documentation on NGDA datasets. We asked agencies to confirm the datasets to verify that the results were accurate. We discussed our findings regarding the search function of the Geospatial Platform with the General Services Administration, which is responsible for the platform catalog's operations. We also conducted a search of the Marketplace tab on the Geospatial Platform to determine how many agencies were posting information about efforts to collect or purchase new geospatial data. To assess the progress of FGDC's and federal departments' and agencies' efforts to collaborate with states, we took the following actions. Our results are not generalizable to all states.

- We analyzed position papers, memos, cost estimates, and other documentation on geospatial data developed by states and state representatives.
- We interviewed state geospatial officials from Maryland, Montana, Ohio, South Carolina, and Washington.
- We interviewed personnel from organizational bodies representing states' interests, including NSGIC, the Association of American Geographers, the Cartography and Geographic Information Society, the Coalition of Geospatial Organizations, the Management Association for Private Photogrammetric Surveyors, and the Urban and Regional Information Systems Association.
- We attended the 2014 NSGIC Midyear Meeting in February 2014, which brought together geospatial data experts representing a large

number of state governments and included sessions on challenges and solutions in coordinating geospatial data between state governments and the federal government.

To determine whether selected federal agencies and states invest in potentially duplicative geospatial data, we extracted 1,054 aerial photography and 10,272 address metadata records from the Geospatial Platform. In order to identify duplicative datasets, we compared each metadata record to the others to identify those records with the same location, and then further analyzed those records using other attributes of aerial photography and address datasets, including date, data format, and resolution. We also compared our selected states' address and aerial photography metadata records on NSGIC's inventory[9] to those from the Geospatial Platform in order to identify those with overlapping locations. We then further analyzed those records using other attributes of aerial photography and address datasets to determine if they were duplicative.

Given that several of our selected states and agencies do not or cannot have key datasets on either the Geospatial Platform or NSGIC's inventory and the difficulties with locating the datasets on the Geospatial Platform, we broadened our search to include the data collection efforts. We also performed literature searches on our key data types to identify current and potential aerial photography and address data coordination and collection efforts at the federal and state levels. We reviewed agency documentation, and reports and whitepapers from sources such as the National Geospatial Advisory Committee, Census, NSGIC, and Interior on aerial photography and address collection. We compared these coordination and collection efforts to each other and to Circular A-16 to identify any duplication or overlap of effort. In addition, we asked selected agency and state officials to identify their key address and aerial photography datasets, and compared those to each other using various attributes to determine the existence of duplication. We reviewed a range of documentation on the key datasets, including standards and specifications. In addition, we interviewed agency and state officials to determine the extent of duplication and the challenges in avoiding duplication in geospatial data. We conducted this performance audit from January 2014 to February 2015 in accordance with generally accepted government auditing standards. Those standards require that we plan and perform the audit to obtain sufficient, appropriate evidence to provide a reasonable basis for our findings and conclusions based on our audit objectives. We believe that the evidence obtained provides a reasonable basis for our findings and conclusions based on our audit objectives.

APPENDIX III: STATUS OF PRIOR GAO RECOMMENDATIONS

The following table provides the implementation status of recommendations from our previous report on geospatial data.[1]

Table 10. Implementation Status for Recommendations from GAO-13-94

Recommendation	FGDC	Commerce	Interior	Transportation	OMB
1. The Secretary of the Interior, as the Federal Geographic Data Committee (FGDC) Chair, should direct the FGDC Steering Committee to establish a time frame for completing a plan to facilitate the implementation of the Office of Management and Budget's (OMB) portfolio management guidance, and develop and implement the plan within the established time frame. The plan, at a minimum, should include goals and performance measures, and the FGDC should report annually to OMB on the progress made on efforts to improve coordination and reduce duplication among themes.	◐	n/a[a]	n/a[a]	n/a[a]	n/a[a]
2. The Secretary of the Interior, as the FGDC Chair, should direct the FGDC Steering Committee to develop and implement guidance for identifying planned geospatial investments using the Geospatial Platform, and establish a time frame for doing so.	●	n/a[a]	n/a[a]	n/a[a]	n/a[a]

Imagery	Georeferenced Images of the earth's surface, which have been collected via aerial photography or satellite data. Orthoimagery is prepared through a geometric correction process known as orthorectification to remove image displacements due to relief and sensor characteristics, allowing their use as base maps for digital mapping and analyses in a geographic information system. Specific imagery datasets created through image interpretation and classification, such as a land cover image, can be found under themes specific to the subject matter. Includes imagery such as Landsat, the National Agriculture Imagery Program, and Digital Orthophoto Quarter Quadrangles.
Land Use-Land Cover	"Land Use-Land Cover" refers collectively to natural and man-made surface features that cover the land (Land Cover) and to the primary ways in which land cover is used by humans (Land Use). Examples of Land Cover may be grass, asphalt, trees, bare ground, and water. Examples of Land Use include urban, agricultural, ranges, and forest areas.
Real Property	The spatial representation (location) of real property entities, typically consisting of one or more of the following: unimproved land, a building, a structure, site improvements and the underlying land. Complex real property entities (that is "facilities") are used for a broad spectrum of functions or missions. This theme focuses on spatial representation of real property assets only and does not seek to describe special purpose functions of real property such as those found in the Cultural Resources, Transportation, or Utilities themes.
Soils	Depicts the geography and attributes of the many kinds of soils found in the landscape at both large and small map scales. A living dynamic resource providing a natural medium for plant growth and habitat for living organisms, soil recycles nutrients and wastes, stores carbon, and purifies water supplies. Soil has distinct layers (called "horizons") that, in contrast to underlying geologic material, are altered by the interactions of climate, landscape features, and living organisms over time.
Transportation	Means and aids for conveying persons and/or goods. The transportation system includes both physical and nonphysical components related to all modes of travel that allow the movement of goods and people between locations.
Utilities	Means, aids, and usage of facilities for producing, conveying, distributing, processing or disposing of public and private commodities including power, energy, communications, natural gas, and water. Includes sub themes Energy, Drinking Water and Water Treatment, and Communications.
Water–Inland	Interior hydrologic features and characteristics, including classification, measurements, location, and extent. Includes aquifers, watersheds, wetlands, navigation, water quality, water quantity, and groundwater information.
Water–Oceans and Coasts	Features and characteristics of salt water bodies (e.g., tides, tidal waves, coastal information, and reefs) and features and characteristics that represent the intersection of the land with the water surface (i.e. shorelines), the lines from which the territorial sea and other maritime zones are measured (i.e., baseline maritime) and lands covered by water at any stage of the tide (e.g., Outer Continental Shelf), as distinguished from tidelands, which are attached to the mainland or an island and cover and uncover with the tide.

Source: Federal Geographic Data Committee. | GAO-15-193

Table 9. (Continued)

Elevation	The measured vertical position of the earth's surface and other landscape or bathymetric features relative to a reference datum typically related to sea level. These points normally describe bare earth positions but may also describe the top surface of buildings and other objects, vegetation structure, or submerged objects. Elevation data can be stored as a three-dimensional array or as a continuous surface such as a raster, triangulated irregular network, or contours. Elevation data may also be represented in other derivative forms such as slope, aspect, ridge and drainage lines, and shaded relief.
Geodetic Control	Survey control points or other related datasets which are accurately tied to the National Spatial Reference System (the official, common federal system for establishing coordinates for geospatial data that are consistent nationwide). Geodetic control examples include passive geodetic control marks, active geodetic observing systems, data from Global Navigation Satellite Systems (e.g., the Global Positioning System), gravity measurements, and models of the earth's gravity field (geoid).
Geology	Geographically referenced data pertaining to the origin, history, composition, structure, features, and processes of the solid earth, both onshore and offshore. Includes geologic, geophysical, and geochemical maps, stratigraphy, paleontology, geochronology, mineral and energy resources, and natural hazards such as earthquakes, volcanic eruptions, coastal erosion, and landslides. Does not include soils.
Governmental Units and Administrative and Statistical Boundaries	Boundaries that delineate geographic areas for uses such as governance and the general provision of services (e.g., states, American Indian reservations, counties, cities, and towns), administration and/or for a specific purpose (e.g., congressional districts, school districts, fire districts, and Alaska Native Regional Corporations), and/or provision of statistical data (census tracts, census blocks, and metropolitan and micropolitan statistical areas). Boundaries for these various types of geographic areas are either defined through a documented legal description or through criteria and guidelines. Other boundaries may include international limits, those of federal land ownership, the extent of administrative regions for various federal agencies, as well as the jurisdictional offshore limits of U.S. sovereignty. Boundaries associated solely with natural resources and/or cultural entities are excluded from this theme and are included in the appropriate subject themes.

Appendix II: Description of National Geospatial Data Asset Data Themes

The table below provides a description of the National Geospatial Data Asset data themes.

Table 9. Definitions of 16 National Geospatial Data Asset Data Themes

Category of data	Definition
Biota	Pertain to, or describe, the dynamic processes, interactions, distributions, and relationships between and among organisms and their environments.
Cadastre	Past, current, and future rights and interests in real property, including the spatial information necessary to describe geographic extents. Rights and interests are benefits or enjoyment in real property that can be conveyed, transferred, or otherwise allocated to another for economic remuneration. Rights and interests are recorded in land record documents. The spatial information necessary to describe geographic extents includes surveys and legal description frameworks such as the Public Land Survey System, as well as parcel-by-parcel surveys and descriptions. Does not include federal government or military facilities.
Climate and Weather	Meteorological conditions, including temperature, precipitation, and wind, that characteristically prevail in a particular region over a long period of time. Weather is the state of the atmosphere at a given time and place, with respect to variables such as temperature, moisture, wind velocity, and barometric pressure.
Cultural Resources	Features and characteristics of a collection of places of significance in history, architecture, engineering, or society. Includes National Monuments and Icons.

3. The Secretary of the Interior, as the FGDC Chair, should direct the FGDC Steering Committee to establish a time frame for creating and updating a strategic plan to improve coordination and reduce duplication, and create and implement the plan within the established time frame. The plan, at a minimum, should include (1) a vision statement for the National Spatial Data Infrastructure (NSDI); (2) outcome-oriented goals and objectives that address all aspects of the NSDI; (3) a description of how the goals and objectives are to be achieved, including a description of the resources needed to achieve the goals and objectives and how the FGDC is to work with other agencies to achieve them; (4) performance measures for achieving the stated goals; and (5) external factors that could affect the achievement of the goals and objectives.	◐	n/a[a]	n/a[a]	n/a[a]	n/a[a]
4. The Secretaries of Commerce and Transportation should designate a senior agency official who has department-wide responsibility, accountability, and authority for geospatial information issues.	n/a[a]	●	n/a[b]	●	n/a[a]

Table 10. (Continued)

5. The Secretaries of the departments in our review should direct their senior agency officials for geospatial information to prepare, maintain, publish, and implement strategies for each of their departments for advancing geographic information and related geospatial data activities appropriate to their missions.	n/a[a]	◐	●	◐	n/a[a]
6. The Secretaries of the departments in our review should direct their senior agency officials for geospatial information to develop policies that require each of the departments to make its geospatial metadata available on the NSDI clearinghouse.	n/a[a]	●	●	●	n/a[a]
7. The Secretaries of the departments in our review should direct their senior agency officials for geospatial information to develop and implement internal procedures to ensure that the departments access the NSDI clearinghouse before they expend funds to collect or produce new geospatial data to determine (1) whether the information has already been collected by others and (2) whether cooperative efforts to obtain the data are possible.	n/a[a]	●	●	●	n/a[a]
8. The Secretaries of the Interior and Transportation should direct their theme points of contact to prepare goals relating to all datasets within the relevant theme that support the NSDI, and as needed, collect and analyze user needs and include those needs in the theme-related goals.	n/a[a]	n/a[b]	●	◐	n/a[a]

9. The Secretaries of the Interior and Transportation should direct their theme points of contact to develop and implement plans for the nationwide population of the relevant themes that addresses all datasets within the theme and that includes (1) the development of partnership programs with states, tribes, academia, the private sector, other federal agencies, and localities that meet the needs of users; (2) human and financial resource needs; (3) standards, metadata, and the clearinghouse needs; and (4) a timetable for the development for the theme.	n/a[a]	n/a[b]	◐	◐	n/a[a]
10. The Secretaries of the departments in our review should direct their theme points of contact to create and implement a plan to develop and implement relevant theme standards.	n/a[a]	●	◐	◐	n/a[a]
11. The Director of the Office of Management and Budget should develop a mechanism, or modify existing mechanisms, to identify and report annually on all geospatial-related investments, including dollars invested and the nature of the investment.	n/a[a]	n/a[a]	n/a[a]	n/a[a]	◐

Key:

= Fully implemented—the agency provided evidence that the recommendation was implemented.

C= Partially implemented—the agency provided evidence that showed progress, but the recommendation was not yet fully implemented.

= Not implemented—the agency did not provide evidence that the recommendation was implemented.

Source: GAO analysis of agency documentation. | GAO-15-193

Notes:

[a]The recommendation was not applicable to that agency because it was outside of the agency's responsibilities.

[b]The recommendation was not applicable because the department or agency was reported by GAO to have been in compliance with the criteria associated with the recommendation at the time of the review.

Acronyms and abbreviations: BLM = Bureau of Land Management; FSA = Farm Service Agency; NOAA = National Oceanic and Atmospheric Administration; USGS = U.S. Geological Survey.

APPENDIX IV: USE OF GEOSPATIAL DATA THEMES BY SELECTED FEDERAL AGENCIES AND STATES

Selected federal agencies and states use a variety of geospatial data to support their missions. Table 11 identifies the data themes used by selected federal agencies. Table 12 identifies the data themes used by selected states. For both tables, we have included addresses[1] even though OMB and FGDC do not recognize it as a theme. We did this because one of our areas of focus in this report is address data.

Table 11. Federal Uses of Geospatial Data Themes

Theme	BLM	Census	FEMA	FSA	NOAA	USGS	USPS
Addresses[a]	Yes	Yes	Yes				Yes
Biota	Yes				Yes	Yes	
Cadastre	Yes	Yes	Yes	Yes		Yes	
Climate and Weather	Yes		Yes		Yes	Yes	
Cultural Resources	Yes	Yes				Yes	
Elevation	Yes		Yes	Yes	Yes	Yes	
Geodetic Control	Yes		Yes		Yes	Yes	

Geology	Yes		Yes			Yes	
Governmental Units, and Administrative and Statistical Boundaries	Yes	Yes	Yes	Yes	Yes	Yes	
Imagery[b]	Yes	Yes	Yes	Yes	Yes	Yes	Yes
Land Use-Land Cover	Yes	Yes	Yes		Yes	Yes	
Real Property	Yes		Yes				
Soils	Yes			Yes		Yes	
Transportation	Yes	Yes	Yes		Yes	Yes	Yes
Utilities	Yes		Yes				
Water—Inland	Yes	Yes	Yes		Yes	Yes	
Water—Oceans and Coasts	Yes	Yes	Yes		Yes		

Source: GAO analysis of agency data. | GAO-15-193

Notes:

[a]OMB and FGDC do not recognize addresses as a data theme. [b] The Imagery data theme includes aerial photography.

Table 12. State Uses of Geospatial Data Themes

Theme	Maryland	Montana	Ohio	South Carolina	Washington
Addresses[a]	Yes	Yes	Yes	Yes	Yes
Biota	Yes	Yes	Yes	Yes	Yes
Cadastre	Yes	Yes	Yes	Yes	Yes
Climate and Weather	Yes	Yes	Yes	Yes	Yes
Cultural Resources	Yes	Yes	Yes	Yes	Yes
Elevation	Yes	Yes	Yes	Yes	Yes
Geodetic Control	Yes	Yes	Yes	Yes	Yes
Geology	Yes	Yes	Yes	Yes	Yes
Governmental Units, and Administrative and Statistical Boundaries	Yes	Yes	Yes	Yes	Yes
Imagery[b]	Yes	Yes	Yes	Yes	Yes
Land use-Land Cover	Yes	Yes	Yes	Yes	Yes
Real Property	Yes	Yes	Yes	Yes	Yes
Soils	Yes	Yes	Yes	Yes	Yes
Transportation	Yes	Yes	Yes	Yes	Yes
Utilities	Yes	Yes	Yes	Yes	Yes
Water—Inland	Yes	Yes	Yes	Yes	Yes
Water—Oceans and Coasts	Yes		Yes	Yes	Yes

Source: GAO analysis of state data. | GAO-15-193

Notes:

[a] OMB and FGDC do not recognize addresses as a data theme.

[b] Acronyms and abbreviations: BLM = Bureau of Land Management; FSA = Farm Service Agency; NOAA = National Oceanic and Atmospheric Administration; USGS = U.S. Geological Survey. The Imagery data theme includes aerial photography.

End Notes

[1] Department of the Interior, Geospatial Line of Business Capital Asset Summary (Aug. 14, 2012).

[2] GAO, Geospatial Information: OMB and Agencies Need to Make Coordination a Priority to Reduce Duplication, GAO-13-94 (Washington, D.C.: Nov. 26, 2012).

[3] The geospatial clearinghouse is one feature of the National Spatial Data Infrastructure. It is intended to be a centralized geospatial metadata repository that contains geospatial metadata records from federal agencies, state and local governments, and academic and private sector organizations that can be searched to determine whether needed geospatial data exist and can be shared.

[4] GAO-13-94.

[5] USPS is an establishment of the federal executive branch. For purposes of our report, we are referring to it as a federal agency.

[6] OMB, Circular No. A-16, Coordination of Geographic Information and Related Spatial Data Activities (Aug. 19, 2002); Issuance of OMB Circular A-16 Supplemental Guidance, M-11-03 (Nov. 10, 2010); and Executive Order No. 12906, Coordinating Geographic Data Acquisition and Access: The National Spatial Data Infrastructure, 59 Fed. Reg. 17,671 (Apr. 11, 1994).

[7] GAO, 2013 Annual Report: Actions Needed to Reduce Fragmentation, Overlap and Duplication, and Achieve Other Financial Benefits, GAO-13-279SP (Washington, D.C.: Apr. 9, 2013) and OMB, Circular No. A-16, Coordination of Geographic Information and Related Spatial Data Activities (Aug. 19, 2002); Issuance of OMB Circular A-16 Supplemental Guidance, M-11-03 (Nov. 10, 2010).

[8] The Bureau of the Budget became OMB in 1970. Executive Order No. 11541, Prescribing the Duties of the Office of Management and Budget and the Domestic Council in the Executive Offices of the President, 35 Fed. Reg. 10737 (July 1, 1970).

[9] In addition to providing a clearinghouse, the Geospatial Platform is also envisioned to provide shared and trusted geospatial services and applications for use by government agencies, their partners, and the public.

[10] http://www.geoplatform.gov.

[11] Data.gov is a website designed to increase the ability of the public to easily find, download, and use datasets that are generated and held by the federal government. The Geospatial Platform and Data.gov use the same data catalog.

[12] NGAC includes members from federal, state, local, and tribal governments, the private sector, and academia who provide FGDC with advice and recommendations related to the management of federal and national geospatial programs.

[13] Orthoimagery are images of the earth's surface collected by aerial photography or satellites that have been corrected for distortions or viewing angles. The National Digital Orthoimagery Program is a consortium of federal and state agencies committed to providing national orthoimagery coverage by coordinating requirements and combining funding resources.

[14] The National Digital Elevation Program brings federal and state agencies together to acquire high-quality elevation data for the United States and its territories.

[15] The 32 agency members of the Steering Committee are: Interior, OMB, the Departments of Agriculture, Commerce, Defense, Education, Energy, Health and Human Services, Homeland Security, Housing and Urban Development, Justice, Labor, State, Transportation, the Treasury, and Veterans Affairs; Environmental Protection Agency; Federal Communications Commission (non-voting member); General Services Administration;

Library of Congress; National Aeronautics and Space Administration; National Archives and Records Administration; National Capital Planning Commission (non-voting member); National Science Foundation; Office of Personnel Management; Small Business Administration; Smithsonian Institution; Social Security Administration; Tennessee Valley Authority; U.S. Agency for International Development; U.S. Army Corps of Engineers (non-voting member); and U.S. Nuclear Regulatory Commission.

[16] NGAC was created by the Secretary of the Interior in December 2007 under the Federal Advisory Committee Act.

[17] A major IT investment is a system or an acquisition requiring special management attention because it has significant importance to the mission or function of the government; significant program or policy implications; high executive visibility; high development, operating, or maintenance costs; an unusual funding mechanism; or is defined as major by the agency's capital planning and investment control process.

[18] The Association of American Geographers is a nonprofit scientific and educational society. Its members share interests in the theory, methods, and practice of geography, which they cultivate through annual meetings and scholarly journals, among other practices. More information can be found at http://www.aag.org.

[19] The National Association of State Chief Information Officers represents state chief information officers and information technology executives and managers from state governments across the United States. Its mission is to foster government excellence through quality business practices, information management, and technology policy. More information can be found at http://www.nascio.org/.

[20] The National Association of Counties is a national organization that represents county governments in the United States. Its mission is to promote sound public policies, foster county solutions and innovation, promote intergovernmental and public-private collaboration and provide value-added services to save counties and taxpayers money. More information can be found at http://www.naco.org/Pages/default.aspx.

[21] The Western Governors' Association represents the governors of 19 western states and 3 U.S.-flag islands. The association is an instrument of the governors for bipartisan policy development, information exchange, and collective action on issues of critical importance to the Western United States. More information can be found at http://www.westgov.org /about.

[22] The National League of Cities represents municipalities from across the country. Its goal is to help city leaders make better communities. More information can be found at http://www.nlc.org/about-nlc.

[23] See NGAC, The Need for a National Address Database, A Report Submitted by the National Geospatial Advisory Committee (December 2012).

[24] FGDC considers aerial photography a subset of the imagery data theme. The other subset is satellite photography.

[25] Leaf-on photography refers to images acquired when there is foliage on certain tree and shrub species. Leaf-off photography refers to images acquired when there is no foliage or a reduced amount of foliage on certain tree and shrub species.

[26] GAO-13-94.

[27] Fragmentation refers to those circumstances in which more than one federal agency (or more than one organization within an agency) is involved in the same broad area of national need and opportunities exist to improve service delivery. Overlap occurs when multiple agencies or programs have similar goals, engage in similar activities or strategies to achieve them, or

target similar beneficiaries. Duplication occurs when two or more agencies or programs are engaged in the same activities or provide the same services to the same beneficiaries.

[28] GAO, 2014 Annual Report: Additional Opportunities to Reduce Fragmentation, Overlap, and Duplication and Achieve Other Financial Benefits, GAO-14-343SP (Washington, D.C.: Apr. 8, 2014); 2013 Annual Report: Actions Needed to Reduce Fragmentation, Overlap and Duplication and Achieve Other Financial Benefits, GAO-13-279SP (Washington, D.C.: Apr. 9, 2013); 2012 Annual Report: Opportunities to Reduce Duplication, Overlap and Fragmentation, Achieve Savings, and Enhance Revenue, GAO-12-342SP (Washington, D.C.: Feb. 28, 2012); and Opportunities to Reduce Potential Duplication in Government Programs, Save Tax Dollars, and Enhance Revenue, GAO-11-318SP (Washington, D.C.: Mar. 1, 2011).

[29] Census tracts are small, relatively permanent statistical subdivisions of a county or equivalent entity. The primary purpose of census tracts is to provide a stable set of geographic units for the presentation of statistical data. Census tracts generally have a population size between 1,200 and 8,000 people, with an optimum size of 4,000 people. Census tract boundaries are delineated with the intention of being maintained over a long time so that statistical comparisons can be made from census to census.

[30] The National Map is a collaborative effort among USGS, other federal agencies, states, and local governments that provides core geospatial data about the United States and its territories, similar to the data traditionally found on paper topographic maps. USGS reported that it is currently transitioning to its 3D Elevation Program as the next generation elevation layer. The program was designed based on the 2012 National Enhanced Elevation Assessment, which identified more than 600 requirements for 3D elevation data to address the mission-critical uses of federal agencies, states, local, and tribal governments, and private industries. The legacy products of the National Elevation Dataset are now a part of the 3D Elevation Program data holdings.

[31] This research group is the Division of Medicaid Policy and Research within the University of South Carolina's Institute for Families in Society.

[32] Department of the Interior, Geospatial Line of Business Capital Asset Summary (Aug. 14, 2012).

[33] GAO, Data Transparency: Oversight Needed to Address Underreporting and Inconsistencies on Federal Award Website, GAO-14-476 (Washington, D.C.: June 30, 2014).

[34] GAO, Information Technology: OMB Needs to Improve Its Guidance on IT Investments, GAO-11-826 (Washington, D.C.: Sept. 29, 2011).

[35] Federal Funding Accountability and Transparency Act. Pub. L. No. 109-282, §§ 1 to 4 (Sept. 26, 2006). This act was amended by the Government Funding Transparency Act of 2008, § 6202(a), Title VI, Ch. 2, Supplemental Appropriations Act, 2008, Pub. L. No. 110- 252 (June 30, 2008); 31 U.S.C. § 6101 Note.

[36] GAO-14-476.

[37] U.S. Department of Commerce, U.S. Census Bureau: High-Quality Maps and Accurate Addresses Are Needed to Achieve Census 2020 Cost-Saving Goals, OIG-12-024-1 (May 10, 2012).

[38] U.S. Department of Commerce, U.S. Census Bureau: 2020 Census Planning: Research Delays and Program Management Challenges Threaten Design Innovation, OIG-14-003-A (Dec. 3, 2013).

[39] GAO, NOAA's Observing Systems: Additional Steps Needed to Achieve an Integrated, Cost-Effective Portfolio, GAO-15-96 (Washington, D.C.: Nov. 17, 2014).

[40] GAO-13-94.

[41] This is FGDC's portfolio management plan for NGDA assets as called for in OMB's Issuance of OMB Circular A-16 Supplemental Guidance (Nov. 10, 2010).

[42] Pub. L. No. 113-101, also known as the DATA Act.

[43] GAO, Federal Data Transparency: Effective Implementation of the DATA Act Would Help Address Government-wide Management Challenges and Improve Oversight, GAO-15-241T (Dec. 3, 2014).

[44] OMB, Circular No. A-16, Coordination of Geographic Information and Related Spatial Data Activities (Aug. 19, 2002); Policies for Federal Agency Public Websites, M-05-04 (Dec. 17, 2004); Designation of a Senior Agency Official for Geospatial Information, M-06- 07 (Mar. 3, 2006); Issuance of OMB Circular A-16 Supplemental Guidance, M-11-03 (Nov. 10, 2010); Digital Government: Building a 21st Century Platform to Better Serve the American People (Washington, D.C.: May 23, 2012); and Executive Order No. 12906, Coordinating Geographic Data Acquisition and Access: The National Spatial Data Infrastructure, 59 Fed. Reg. 17,671 (Apr. 11, 1994).

[45] OMB, Circular No. A-16, Coordination of Geographic Information and Related Spatial Data Activities (Aug. 19, 2002); Policies for Federal Agency Public Websites, M-05-04 (Dec. 17, 2004); Designation of a Senior Agency Official for Geospatial Information, M-06- 07 (Mar. 3, 2006); Issuance of OMB Circular A-16 Supplemental Guidance, M-11-03, (Nov. 10, 2010); Digital Government: Building a 21st Century Platform to Better Serve the American People (Washington, D.C.: May 23, 2012); and Executive Order No. 12906, Coordinating Geographic Data Acquisition and Access: The National Spatial Data Infrastructure, 59 Fed. Reg. 17,671 (Apr. 11, 1994).

[46] GAO-13-94.

[47] The Marketplace is a function within the platform for agencies to identify their planned acquisitions to foster cooperative efforts to acquire data.

[48] GAO-13-94.

[49] OMB, Policies for Federal Agency Public Websites, M-05-04 (Dec. 17, 2004); and Digital Government: Building a 21st Century Platform to Better Serve the American People (Washington, D.C.: May 23, 2012).

[50] These estimates are based on our selection of generalizable samples of datasets from the Geospatial Platform catalog and have a margin of error of plus or minus 7 and 4 percentage points respectively. See appendix I for more details on our selection of the generalizable samples.

[51] USPS is not required to register its datasets.

[52] The State Broadband Initiative program's goal was to assist states in gathering data on the availability, speed, and location of broadband services, as well as the broadband services that community institutions, such as schools, libraries, and hospitals, use. These data were used to facilitate the integration of broadband and information technology into state and local economies.

[53] Transportation for the Nation is an ongoing project by the Department of Transportation to develop comprehensive, publicly available, nationwide datasets for all modes of transportation, including roads, railways, airports, and water.

[54] The Coastal Services Center collects coastal LIDAR elevation data from a variety of federal, state, and local governments.

[55] Under the National Digital Elevation Program, federal and state agencies work together to acquire high-quality elevation data for the United States and its territories. USGS reported that it is currently transitioning to its 3D Elevation Program as the next generation elevation layer. The program was designed based on the 2012 National Enhanced Elevation

Assessment, which identified more than 600 requirements for 3D elevation data to address the mission-critical uses of federal agencies, states, local, and tribal governments, and private industries.

[56] The National Wetlands Inventory was established to conduct a nationwide inventory of U.S. wetlands to provide biologists and others with information on the distribution and type of wetlands to aid in conservation efforts.

[57] Although the program is administered by the National Geospatial-Intelligence Agency, to date the data collection, management, handing, and quality control aspects of the program have been executed by the USGS.

[58] 13 U.S.C. § 9 prohibits Census from disclosing data it collects about individuals and establishments in a manner that would identify those individuals and businesses.

[59] The National Agriculture Imagery Program acquires 1-meter aerial imagery during the agricultural growing seasons in the coterminous United States in a 2-year cycle. The resulting digital orthophotography is available to governmental agencies and the public for a nominal fee. FSA also allows federal, state, and local agencies to contract for more detailed orthophotography collections, known as "buy-ups," subject to its approval, and at additional costs to the partners.

[60] OMB, Circular No. A-16, Coordination of Geographic Information and Related Spatial Data Activities (Aug. 19, 2002).

[61] A 2011 study for the Census Bureau showed that the majority of private vendors rely primarily on USPS for their address data. See M. Dobson, Ph.D., D. Cowen, Ph.D., and S. Guptill, Ph.D. Reporting the State and Anticipated Future Directions of Addresses and Addressing, a report to the Geography Division, U.S. Bureau of the Census, Deliverable #2 – Syneren Technologies Contract – Task T005 (Jan. 5, 2011).

[62] NGAC, The Need for a National Address Database, A Report Submitted by the National Geospatial Advisory Committee (December 2012).

[63] 13 U.S.C. § 9 prohibits Census from disclosing data it collects about individuals and establishments in a manner that would identify those individuals and businesses. Similarly, 39 U.S.C. § 412 prohibits USPS from disclosing names or addresses (past or present) of postal patrons or other persons. However, the Census Address List Improvement Act of 1994 authorized USPS to share its address information with Census.

[64] Personally identifiable information is any information that can be used to distinguish or trace an individual's identity, such as name, date, and place of birth, Social Security number, or other types of personal information that can be linked to an individual, such as medical, educational, financial, and employment information.

[65] GAO, 2020 Census: Census Bureau Can Improve Use of Leading Practices When Choosing Address and Mapping Sources, GAO-15-21 (Washington, D.C.: Oct. 2, 2014).

[66] GAO, Critical Infrastructure Protection: More Comprehensive Planning Would Enhance the Cybersecurity of Public Safety Entities' Emerging Technology, GAO-14-125 (Washington, D.C.: Jan. 28, 2014).

[67] The Urban and Regional Information Systems Association is a nonprofit association of professionals using GIS and other information technologies to solve challenges in state/provincial, regional, and local government agencies and departments.

[68] These agencies consist of six Interior agencies (USGS, BLM, Bureau of Reclamation, Bureau of Indian Affairs, National Park Service, and U.S. Fish and Wildlife Service) and USDA's Natural Resources Conservation Service and U.S. Forest Service.

[69] Pub. L. No. 113-101, also known as the DATA Act.

End Notes for Appendix I

[1] The geospatial clearinghouse is one feature of the National Spatial Data Infrastructure. It is intended to be a centralized geospatial metadata repository that contains geospatial metadata records from federal agencies, state and local governments, and academic and private sector organizations that can be searched to determine whether needed geospatial data exist and can be shared. The clearinghouse is currently part of the Federal Geographic Data Committee's Geospatial Platform.

[2] GAO, Geospatial Information: OMB and Agencies Need to Make Coordination a Priority to Reduce Duplication, GAO-13-94 (Washington, D.C.: Nov. 26, 2012).

[3] As of December 2014, only BLM did not participate in the Geospatial Platform.

[4] GAO, Data Transparency: Oversight Needed to Address Underreporting and Inconsistencies on Federal Award Website, GAO-14-476 (Washington, D.C.: June 30, 2014); Information Technology: Reform Initiatives Can Help Improve Efficiency and Effectiveness, GAO-14-671T (Washington, D.C.: June 10, 2014); IT Dashboard: Agencies Are Managing Investment Risk, but Related Ratings Need to Be More Accurate and Available, GAO-14-64 (Washington, D.C.: Dec. 12, 2013); Information Technology Dashboard: Opportunities Exist to Improve Transparency and Oversight of Investment Risk at Select Agencies, GAO-13-98 (Washington, D.C.: Oct. 16, 2012); IT Dashboard: Accuracy Has Improved, and Additional Efforts are Under Way to Better Inform Decision Making, GAO-12-210 (Washington, D.C.: Nov. 7, 2011); Information Technology: OMB Has Made Improvements to Its Dashboard, but Further Work is Needed by Agencies and OMB to Ensure Data Accuracy, GAO-11-262 (Washington, D.C.: Mar. 15, 2011); and Information Technology: OMB's Dashboard Has Increased Transparency and Oversight, but Improvements Needed, GAO-10-701 (Washington, D.C.: July 16, 2010).

[5] Executive Order No. 12906, Coordinating Geographic Data Acquisition and Access: The National Spatial Data Infrastructure, 59 Fed. Reg. 17,671 (Apr. 11, 1994).

[6] OMB, Circular No. A-16, Coordination of Geographic Information and Related Spatial Data Activities (Aug. 19, 2002).

[7] OMB, Issuance of OMB Circular A-16 Supplemental Guidance, M-11-03 (Nov. 10, 2010).

[8] GAO-13-94.

[9] http://gisinventory.net.

End Note for Appendix III

[1] GAO, Geospatial Information: OMB and Agencies Need to Make Coordination a Priority to Reduce Duplication, GAO-13-94 (Washington, D.C.: Nov. 26, 2012).

End Note for Appendix IV

[1] Address data include the house number and street or road name or other designation assigned to a housing unit, special place, business establishment, or other structure for purposes of mail delivery and/or to enable emergency services, delivery people, and visitors to find the structure. Sometimes addresses include a geocoded reference such as latitude and longitude coordinates.

In: Geospatial Data
Editors: Meredith Haynes

ISBN: 978-1-63483-524-4

Chapter 2

GEOSPATIAL INFORMATION: OMB AND AGENCIESNEED TO MAKE COORDINATION A PRIORITY TO REDUCE DUPLICATION*

United States Government Accountability Office

ABBREVIATIONS

BTS	Bureau of Transportation Statistics
Commerce	Department of Commerce
FGDC	Federal Geographic Data Committee
GIS	geographic information systems
Interior	Department of the Interior
IT	information technology
NOAA	National Oceanic and Atmospheric Administration
NSDI	National Spatial Data Infrastructure
OMB	Office of Management and Budget
TIGER	Topologically Integrated Geographic Encoding and Referencing

* This is an edited, reformatted and augmented version of a Congressional Research Service publication GAO-13-94, prepared for Members and Committees of Congress dated November 2012.

Transportation Department of Transportation
USGS U.S. Geological Survey

WHY GAO DID THIS STUDY

The federal government collects, maintains, and uses geospatial information—information linked to specific geographic locations—to support many functions, including national security and disaster response. In 2012, the Department of the Interior (Interior) estimated that the federal government invests billions of dollars on geospatial data annually, and that duplication is common. GAO was asked to determine the extent to which the federal government has established and effectively implemented policies and procedures for coordinating its geospatial investments and avoiding duplication.

To do so, GAO focused on FGDC coordination activities; efforts within the departments of Commerce, the Interior, and Transportation; and OMB oversight. GAO reviewed FGDC and department documentation, such as policies, procedures, and strategic plans; OMB guidance and an executive order; and reports concerning duplicative investments.

WHAT GAO RECOMMENDS

GAO is making recommendations aimed at improving coordination and reducing duplication, to include FGDC developing a national strategy for coordinating geospatial investments; federal agencies following federal guidance for managing geospatial investments; and OMB developing a mechanism to identify and report on geospatial investments. Two agencies and OMB generally agreed with GAO's recommendations and one agency neither agreed nor disagreed.

WHAT GAO FOUND

While the President and the Office of Management and Budget (OMB) have established policies and procedures for coordinating investments in geospatial data, governmentwide committees and federal departments and agencies have not effectively implemented them. The committee that was established to promote the coordination of geospatial data nationwide—the Federal Geographic Data Committee (FGDC)—has developed and endorsed key standards— including a metadata standard that includes descriptive information about a particular set of geospatial data—and established a clearinghouse of metadata; however, the clearinghouse is not being used by agencies to identify planned geospatial investments to promote coordination and reduce duplication. The FGDC has not yet planned or implemented an approach to manage geospatial data as related groups of investments to allow agencies to more effectively plan geospatial data collection efforts and minimize duplicative investments; and its strategic plan is missing key elements, such as performance measures for many of its defined objectives. Further, none of the three federal departments in GAO's review have fully implemented important activities for coordinating geospatial data, such as preparing and implementing a strategy for advancing geospatial activities within their respective departments (see table).

Further, the three agencies in GAO's review responsible for governmentwide management of specific geospatial data have implemented some but not all important activities for coordinating the national coverage of specific geospatial data. For example, only one agency has developed a plan for the nationwide population of the datasets under its responsibility, and none of the agencies have developed a plan to develop standards that facilitate the collection and sharing of geospatial data. Finally, while OMB has oversight responsibilities for geospatial data, OMB staff members acknowledged that OMB does not have complete and reliable information to identify potentially duplicative geospatial investments.

FGDC, federal departments and agencies, and OMB have not yet fully implemented policies and procedures for coordinating geospatial investments because these efforts have not been a priority. As a result, efforts to acquire data are uncoordinated and the federal government is acquiring duplicative geospatial data. For example, three agencies are independently acquiring road data, which is reported to have resulted in millions of wasted taxpayers' dollars. Unless OMB, the FGDC, and federal departments and agencies decide that coordinating geospatial investments is a priority, this situation will likely continue.

November 26, 2012

The Honorable Joseph I. Lieberman
Chairman
The Honorable Susan M. Collins
Ranking Member
Committee on Homeland Security and Governmental Affairs
United States Senate

The federal government collects, maintains, and uses geospatial information—information linked to specific geographic locations—to help in decision making and to support many functions, including national security, law enforcement, health care, environmental protection, and natural resources conservation. Among the many activities that can depend on critical analysis of geospatial information are maintaining roads and other critical transportation infrastructure and quickly responding to natural disasters, such as floods, hurricanes, and fires.

Multiple federal agencies may provide services at the same geographic locations and may independently collect similar geospatial information about those locations, thus raising the question of how well the nation's investments in geospatial data are coordinated. Moreover, the Department of the Interior (Interior) has recently estimated that the federal government invests billions of dollars in geospatial data annually, and reported that duplication among investments is common.[1]

Over the past 2 years, we issued two comprehensive reports that identified federal programs or functional areas where duplication,[2] overlap, or fragmentation exists; the actions needed to address such conditions; and the potential financial and other benefits of doing so.[3] For this review, you asked us to determine the extent to which the federal government has established and effectively implemented policies and procedures for coordinating its investments in geospatial data and avoiding duplication.

To address our objective, we focused on governmentwide activities to implement the National Spatial Data Infrastructure (NSDI)—an infrastructure to facilitate the efficient collection, sharing, and dissemination of geospatial data among all levels of government, and public and private sectors—as well as efforts of the Federal Geographic Data Committee (FGDC)—the federal committee established to promote the coordinated use, sharing, and dissemination of geospatial data nationwide. Additionally, we focused on activities within three selected departments: Department of Commerce

(Commerce), Department of the Interior (Interior), and Department of Transportation (Transportation); and within three selected agencies responsible for managing data themes: 4 the National Oceanic and Atmospheric Administration (NOAA), the U.S. Geological Survey (USGS), and the Bureau of Transportation Statistics (BTS). The themes in our review are geodetic control[5]—(NOAA), hydrography[6]—(USGS), and transportation[7]—(BTS). We reviewed and assessed FGDC and department documentation, such as policies, procedures, strategic plans, meeting minutes, and budget documentation; the Office of Management and Budget (OMB) data call results and guidance; and recent reports discussing duplicative geospatial investments; and interviewed FGDC and department officials, and staff members from the OMB Office of E-Government.

We conducted this performance audit from November 2011 to November 2012, in accordance with generally accepted government auditing standards. Those standards require that we plan and perform the audit to obtain sufficient, appropriate evidence to provide a reasonable basis for our findings and conclusions based on our audit objective. We believe that the evidence obtained provides a reasonable basis for our findings and conclusions based on our audit objective. Details of our objective, scope, and methodology are contained in appendix I.

BACKGROUND

Geospatial information describes entities or phenomena that can be referenced to specific locations relative to the Earth's surface. For example, entities, such as buildings, rivers, road intersections, power plants, and national parks can all be identified by their locations. In addition, phenomena, such as wildfires, the spread of the West Nile virus, and the thinning of trees due to acid rain can also be identified by their geographic locations.

Users can analyze that data in geographic information systems (GIS)—systems of computer software, hardware, and data used to capture, store, manipulate, analyze, and graphically present a potentially wide array of geospatial information. The primary function of a GIS is to link multiple sets of geospatial data and display the combined information as maps with different layers of information. Assuming that all of the information is at the same scale and has been formatted according to the same geospatial standards, users can potentially overlay geospatial information about any number of specific topics to examine how the data in the various layers interrelate. Each layer of a GIS map typically represents a single theme comprised of one or more sets of data, each of which could be derived from a source completely different from the others. Figure 1 portrays the concept of visual representation of geospatial data themes in a GIS.

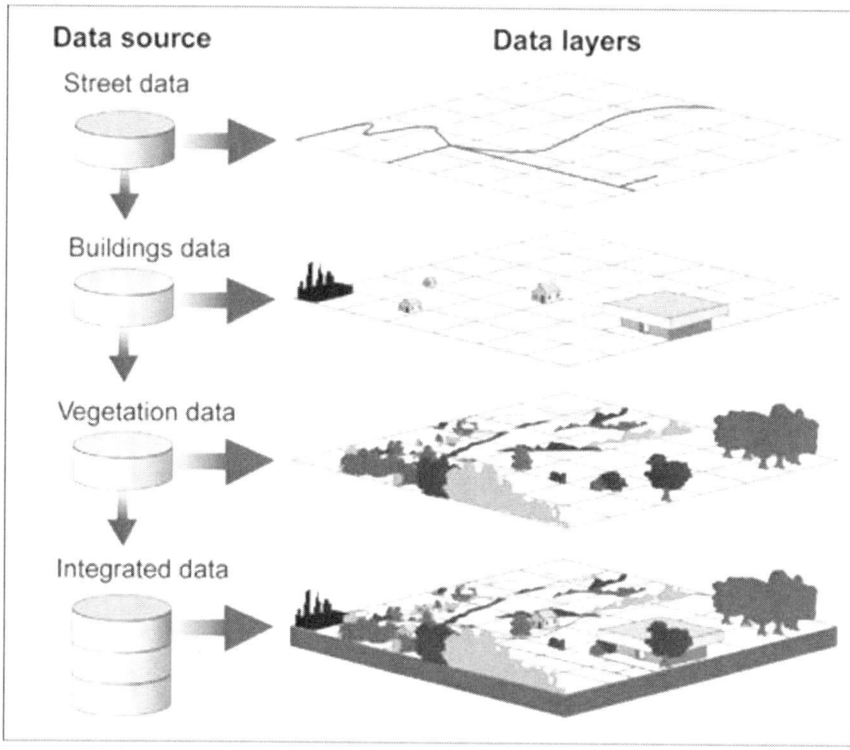

Source: GAO.

Figure 1. Visual Representation of Themes in a GIS.

Examples of geospatial data applications are provided in figures 2 and 3. Figure 2 demonstrates the usefulness of GIS in showing the scope, severity, and duration of the effects of the recent drought in the United States, which could be used to make drought relief and agricultural support activities more effective.

Another use for GIS is tracking and responding to natural disasters, such as wildfires and hurricanes. Figure 3 demonstrates the usefulness of GIS in tracking the direction and estimated strength of an impending hurricane. The timely delivery of these data can be used to provide for orderly evacuation of people from affected areas, and lessen the impact of the storm on facilities, such as sewage treatment plants, hospitals, and nursing homes.

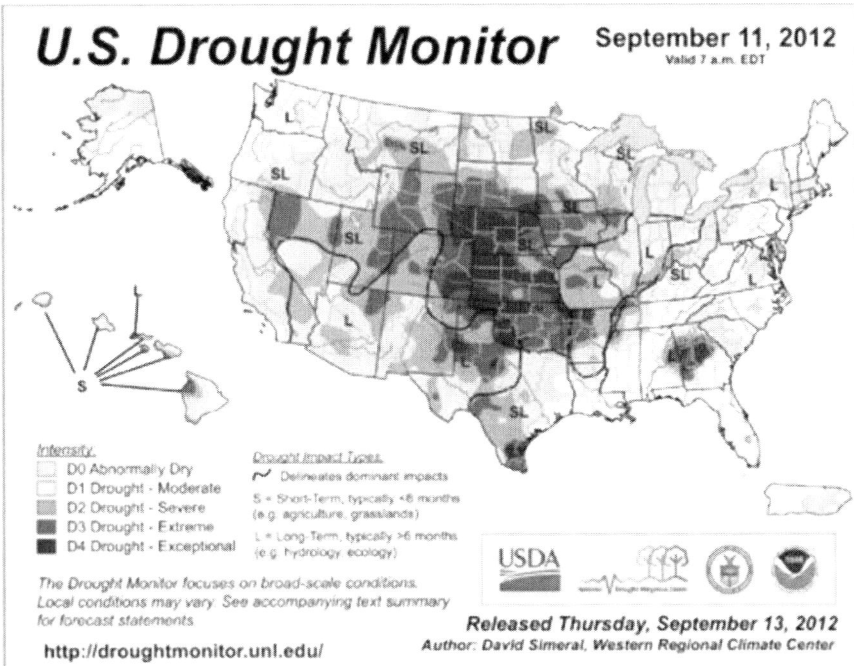

Source: The National Drought Mitigation Center at the University of Nebraska-Lincoln. The U.S. Drought Monitor is a partnership between the National Drought Mitigation Center at the University of Nebraska-Lincoln, the United States Departmwnt of Agriculture and the National Oceanic and Atmospheric Administration.

Figure 2. Monitoring Drought Conditions across the United States.

Coordination of Federal Geospatial Activities

For many years, the federal government has taken steps to coordinate geospatial activities both within and outside the federal government. In 1953, the Bureau of the Budget[8] first issued Circular A-16, encouraging expeditious surveying and mapping activities across all levels of government and avoidance of duplicative efforts. In 1990, OMB revised Circular A-16 to, among other things, establish the FGDC within Interior to promote the coordinated use, sharing, and dissemination of geospatial data nationwide.

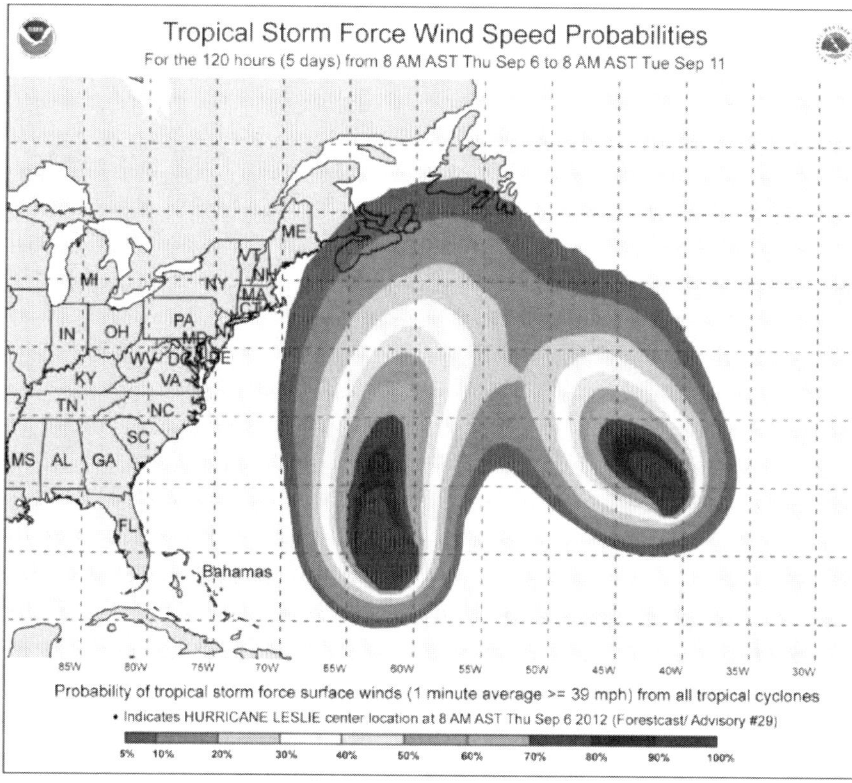

Source: National Oceanic and Atmospheric Administration.

Figure 3. Hurricane Leslie Wind Speed Predictions.

Building on that guidance, in 1994 the President issued Executive Order 12906 for the purpose of addressing wasteful duplication and incompatibility of geospatial information, and assigned the FGDC the responsibility to coordinate the development of the NSDI.[9] In 2002, OMB again revised Circular A-16 to further describe the components of the NSDI; clearly define agency responsibilities for acquiring, maintaining, distributing, using, and preserving geospatial data; and to reaffirm the FGDC's role as the interagency coordinating body for NSDI-related activities.[10] The circular established the following five components of the NSDI and described how these components were to be implemented.

- *Data themes.* Data themes are topics of national significance, such as transportation, which includes all modes of travel (e.g., road and rail data). OMB Circular A-16 currently identifies 34 data themes and identifies the "lead" agency or agencies for each theme. Each data theme is to be comprised of one or more electronic data records, known as datasets. Of the 34 themes, 9 are identified as "framework" themes 11 —that is, themes identified in Circular A-16 as critical for many geospatial applications.

- *Standards.* Geospatial standards provide common and repeatable rules or guidelines for the development, documentation, and exchange of geospatial datasets.
- *Metadata.* Metadata are information about datasets, such as content, source, accuracy, method of collection, and point-of-contact. Metadata are used to facilitate the search of and access to datasets within a data library or clearinghouse, and enable potential users to determine the data's applicability for their use.
- *National Spatial Data Clearinghouse.* The clearinghouse is intended to be a centralized geospatial metadata repository that contains geospatial metadata records from federal agencies, state and local governments, and academic and private sector organizations that can be searched to determine whether needed geospatial data exist and can be shared. Federal agencies are required to identify their existing and planned geospatial investments in the clearinghouse, and search the clearinghouse for cost-saving opportunities before acquiring geospatial data. In 2003, the FGDC created the Geospatial One-Stop to provide "one-stop" access to geospatial metadata from a centralized database and search function. In October 2011, the Geospatial One-Stop was retired, and the FGDC initiated a pilot project, known as the Geospatial Platform, which is envisioned to provide shared and trusted geospatial data, services, and applications for use by government agencies, their partners, and the public.[12] According to Interior officials, Interior is the managing partner of the Geospatial Platform. As of August 2012, there were approximately 835,000 geospatial metadata records in the central repository, of which about 373,000 were from federal sources. Users can search the metadata repository through two primary portals: geo.data.gov[13] and the Geospatial Platform.[14] The General Services Administration is responsible for managing the clearinghouse database and the associated web portal geo.data.gov.
- *Partnerships.* Partnerships are efforts aimed at involving all stakeholders (e.g., federal, tribal, state, local government, and academic institutions) in the development of the NSDI.

In November 2010, OMB issued supplemental guidance specifically regarding how agencies are to manage data themes.[15] This supplemental guidance expands upon and clarifies some of the language and responsibilities contained in OMB Circular A-16 in order to facilitate the adoption and implementation of a geospatial asset management capability.

To fulfill its responsibilities, the FGDC is governed by a steering committee—an interagency decision making body that provides leadership and policy direction in support of the development of the NSDI. The Secretary of the Interior chairs the committee; the Vice-Chair is the Deputy Director for Management of OMB.[16] All departments or agencies responsible for geospatial data themes or that have activities in geographic information or geospatial data collection or use are required to be members of the FGDC. Thirty-two agencies[17] are currently members of the Steering Committee and are to be represented by their senior agency officials for geospatial information.[18] These senior agency officials are responsible for overseeing, coordinating, and facilitating their respective agency's implementation of geospatial requirements, policies, and activities. The FGDC is supported by an Office of the Secretariat that consists of about 10 people located in USGS who do the day-to-day work of supporting, managing, and coordinating the activities of the FGDC.

In addition, in December 2007, the Secretary of the Interior created the National Geospatial Advisory Committee [19] to provide the department and the FGDC with advice and recommendations related to the management of federal and national geospatial programs, development of the NSDI, and the implementation of related federal guidance. Members of the committee include approximately 30 officials from federal, state, local, and tribal governments, the private sector, and academia.

OMB's Roles and Responsibilities for Overseeing IT Investments

OMB has specific oversight responsibilities for federal information technology (IT) systems and acquisition activities—including GIS—to help ensure their efficient and effective use. Two key laws that outline these responsibilities are the Clinger-Cohen Act of 1996[20] and the E-Government Act of 2002.[21]

- The Clinger-Cohen Act of 1996—The act requires OMB to establish processes to analyze, track, and evaluate the risks and results of major capital investments in information systems made by federal agencies and report to Congress on the net program performance benefits achieved as a result of these investments.
- The E-Government Act of 2002—The act establishes an e-government initiative, which encourages the use of web-based Internet applications to enhance the access to and delivery of government information and service to citizens, to business partners, to employees, and among all levels of government. The act also requires OMB to report annually to Congress on the status of e-government initiatives. In these reports, OMB is to describe the administration's use of e-government principles to improve government performance and the delivery of information and services to the public.

OMB subsequently began initiatives to fulfill the requirements established by these laws:

- In February 2002, OMB established the Federal Enterprise Architecture, which is intended to facilitate governmentwide improvement through cross-agency analysis and identification of duplicative investments, gaps, and opportunities for collaboration, interoperability, and integration within and across agency programs. The Federal Enterprise Architecture is composed of five "reference models" describing the federal government's (1) business (or mission) processes and functions, independent of the agencies that perform them; (2) performance goals and outcome measures; (3) means of service delivery; (4) information and data definitions; and (5) technology standards.
- In March 2004, OMB established multiple "lines of business" to consolidate redundant IT investments and business processes across the federal government. Later, in March 2006, OMB established the Geospatial Line of Business. Each line of business is led by an individual agency and supported by other relevant agencies. Interior is the managing partner for the Geospatial Line of Business and the FGDC Secretariat provides project management support. OMB reports to Congress each year on the costs and benefits of these initiatives.

In carrying out its responsibilities, OMB uses several data collection mechanisms to oversee federal IT spending during the annual budget formulation process. Specifically, OMB requires federal departments and agencies to provide information related to their IT investments (called exhibit 53s) and capital asset plans and business cases (called exhibit 300s).

- Exhibit 53. The purpose of the exhibit 53 is to identify all IT investments—both major and nonmajor—and their associated costs within a federal organization. Information included in agency exhibit 53s is designed, in part, to help OMB better understand agencies' spending on IT investments. OMB guidance for the fiscal years 2013 and 2014 budget formulation instructs agencies to identify their geospatial investments in the exhibit 53 using Federal Enterprise Architecture codes for specific functions (e.g., geospatial services, financial management, and acquisition management).
- Exhibit 300. The purpose of the exhibit 300 is to provide a business case for each major IT investment and to allow OMB to monitor IT investments once they are funded. Agencies are required to provide information on each major investment's cost, schedule, and performance.

GAO Has Previously Recommended Improvements in the Management of Geospatial Information and OMB Guidance

In June 2004, we reported that OMB, individual federal agencies, and cross-government committees and initiatives, such as the FGDC and the Geospatial One-Stop project, had taken actions to coordinate the government's geospatial investments across agencies and with state and local governments.[22] However, we noted that these efforts had not been fully successful in reducing duplication in geospatial investments for several reasons:

- a complete and up-to-date strategic plan for doing so was not in place;
- agencies had not consistently complied with OMB guidance that seeks to identify and reduce duplication; and
- OMB's oversight of federal geospatial activities had not been effective because its methods—the annual budget review process, the federal enterprise architecture effort, and the FGDC's reporting

process— were insufficiently developed and had not produced consistent and complete information.

We reported that, as a result of these shortcomings, federal agencies were still independently acquiring and maintaining potentially duplicative and costly datasets and systems. Accordingly, we recommended that the Director of OMB and the Secretary of the Interior direct the development of a national geospatial strategic plan, and recommended that the Director of OMB develop criteria for assessing interagency coordination on proposals for potential geospatial investments, and strengthen its oversight of geospatial projects. OMB and Interior generally agreed with our recommendations. In response, between 2004 and 2008, OMB, Interior, and the FGDC created a number of documents that addressed the development of a national geospatial strategic plan, including a strategic plan for NSDI development and a business case for the development of the Geospatial Line of Business. Furthermore, in 2004 and 2006, OMB issued guidance designed to increase the amount of budget information available on geospatial investments, and improve oversight of agencies' implementation of geospatial-related requirements, policies, and activities.

In September 2011, we reported that OMB's guidance to agencies for reporting their IT investments did not ensure complete reporting or facilitate the identification of duplicative investments.[23] Specifically, agencies differed on what investments they included as an IT investment. We further reported that OMB's guidance to federal agencies to categorize IT investments did not go far enough to allow for the identification of potentially duplicative investments. In particular, OMB's guidance required that each investment be mapped to a single Federal Enterprise Architecture functional code; however, IT investments could fit into more than one functional code.

Accordingly, we recommended that OMB clarify its guidance on reporting IT investments to specify whether certain types of systems—such as space systems—are to be included; allow agencies to place their IT investments into more than one Federal Enterprise Architecture functional code in order to reduce potentially duplicative investments; and direct agencies to report the overall steps that they take to ensure that their IT investments are not duplicative as part of their annual budget and IT investment submissions. OMB did not agree that further efforts were needed to clarify reporting in regard to the types of systems; but it agreed with our recommendations regarding the categorization of investments and reporting of steps taken to reduce duplication. OMB's fiscal year 2014 budget formulation guidance allows agencies to identify up to five Federal Enterprise Architecture functional codes with each investment.

Implementing Established Policies Is Not a Federal Priority, Resulting in Duplicative Investments

While the President and OMB have established policies and procedures for managing and coordinating investments in geospatial data, the FGDC, federal departments and theme-lead agencies, and OMB itself have not effectively implemented them. This has resulted in uncoordinated and duplicative investments in areas of national interest, such as road and address data.

- While the FGDC has developed and endorsed several standards, it has not yet planned for or implemented an approach to manage data themes and their associated key datasets[24] as related groups of investments designed to allow agencies to more effectively plan geospatial data collection efforts and minimize duplicative investments. Additionally, planned geospatial data acquisitions are not identified in the clearinghouse and the FGDC does not have a current strategic plan to guide its efforts.
- None of the three federal departments in our review have fully implemented important activities for coordinating geospatial data and assets, such as developing and implementing a strategy for advancing geospatial activities within the department.
- The three theme-lead agencies in our review have implemented some but not all important activities to ensure the national coverage and stewardship of geospatial data themes.
- OMB's annual budget reporting mechanisms have not provided complete and reliable information to identify duplicative geospatial investments.

The primary cause for why the FGDC, federal departments and theme-lead agencies, and OMB have not yet fully implemented established policies and procedures for coordinating geospatial investments is because, according to OMB staff members and agency officials, they have been focusing on other

priorities. Because federal agencies have yet to fully implement important activities and practices for coordinating and managing geospatial data and facilitating the development of the NSDI, and OMB is limited in its ability to oversee agencies' geospatial investments, agencies continue to acquire duplicative geospatial data.

FGDC Has Developed Geospatial Standards, but Has Not Fully Implemented Key Activities for Coordinating Geospatial Data

According to federal guidance,[25] the FGDC is to serve as the lead federal executive body charged with the leadership, development, implementation, and review of geospatial data standards.

To its credit, the FGDC has developed and endorsed several standards. In particular, it developed a metadata standard that includes descriptive information about a dataset—such as who created and published it, the related theme keyword, and the geographic coordinates that bound the dataset—and facilitates the identification and sharing of geospatial data. 26 The FGDC has also developed standards associated with each of the framework themes and endorsed several other standards developed by external standards bodies, such as the International Organization for Standardization's tracking and navigation standard for web services.[27]

FGDC Has Not Yet Fully Planned or Implemented a Portfolio Management Approach

OMB guidance[28] from November 2010 calls for the immediate use of an approach to manage the NSDI data themes and their associated key datasets as related groups of investments, known as portfolio management. A portfolio management approach establishes a framework for governmentwide management of themes and datasets to allow agencies to more effectively plan geospatial data collection efforts and minimize duplicative investments. It includes establishing goals and performance measures, as well as processes for reviewing the health and status of datasets across the government in order to maximize the value of the data. OMB further directs the FGDC to provide guidance, within a year of issuance, about how to implement the portfolio management approach.

While the FGDC has initiated activities that Secretariat officials say are first needed to establish a portfolio of datasets, it has not yet fully planned for or implemented a portfolio management approach. Specifically, the FGDC evaluated the 34 data themes identified in OMB Circular A-16 to determine whether any changes were needed; in August 2011, the Steering Committee proposed consolidating the 34 data themes into 17 themes; Secretariat officials subsequently stated that the FGDC agencies are proposing to eliminate 1 more theme for a total of 16. (See app. II for a comparison of the 34 themes and the newly proposed 16 themes.) These officials further stated that, as of August 2012, lead agencies have been identified for each of the 16 themes and said that they plan to discuss the revised lists of themes and lead agencies at the Steering Committee's September 2012 meeting. Once the Steering Committee approves the revised themes and lead agencies, the FGDC plans to send them to OMB for its approval. Additionally, the FGDC has identified 221 key datasets associated with the proposed data themes.[29]

However, the data themes, lead agencies, and datasets have neither been finalized nor approved, and the FGDC has yet to provide guidance to agencies about how to implement the portfolio management approach. While Secretariat officials stated that they had developed a draft implementation plan in November 2011, it has not been finalized or approved, and FGDC Secretariat officials were unable, on behalf of FGDC agencies, to provide a time frame for doing so.

FGDC Secretariat officials stated that completion of the activities needed to fully implement the portfolio management approach has not been accomplished due to competing priorities, such as the Geospatial Platform. Until the implementation plan is completed; and the themes, lead agencies, and associated datasets are identified and approved, the agencies cannot effectively begin to implement a coordinated geospatial asset management capability that was, according to the OMB guidance, designed to provide a mechanism for agencies to plan more effectively in advance of data collection efforts to identify partnership opportunities, and to minimize duplicative investments.

FGDC Has Established a Clearinghouse, but Planned Acquisitions Are Not Identified and Shared

According to federal guidance,[30] the FGDC is to develop a clearinghouse with the functionality to allow federal departments and agencies to (1) determine whcther the geospatial data they are seeking already exist and (2) identify planned acquisitions of geospatial data and opportunities to jointly acquire the data in order to improve efficiencies in geospatial data collection and to reduce potential redundancies and duplication. Additionally, federal guidance requires agencies to identify their planned investments in the clearinghouse.[31]

The FGDC has developed a clearinghouse that allows users to determine whether the data they are seeking exist. As noted previously, the clearinghouse consists of a centralized repository that contains geospatial metadata[32] records from federal agencies, state and local governments, and academic and private-sector organizations; and multiple web-based portals from which the metadata can be searched. The two primary portals are geo.data.gov and the Geospatial Platform. As of August 2012, there were approximately 835,000 geospatial metadata records in the centralized repository, of which about 373,000 were from federal sources.

Although the clearinghouse allows agencies to identify their planned investments, federal agencies are not doing so. According to the Geospatial Platform managing partner representative, the platform was modified in May 2012 to include a site for agencies to identify their planned acquisitions and potential cooperative efforts to acquire data. However, as of September 2012, federal agencies have not identified any planned geospatial investments. The Geospatial Platform managing partner representative stated that agencies are not identifying their planned investments because the FGDC has not completed and shared guidance with federal agencies that describes what information is to be shared and how agencies are to identify the planned investments in the platform. This official acknowledged that until the guidance is completed and implemented, agencies are not likely to add their planned acquisitions to the platform and identify potential cooperative efforts to acquire geospatial data and minimize potential redundancies and duplicative efforts.[33]

According to Secretariat officials, they have not yet finalized the guidance for placing planned acquisitions on the platform because the primary focus of the FGDC has been on the development of the Geospatial Platform's Business Plan and the initial release of the Geospatial Platform's core capabilities, applications, and tools. However, without the ability to identify planned geospatial data acquisitions, agencies will likely miss opportunities to reuse or cooperatively acquire geospatial data, thus resulting in the acquisition of potentially duplicative geospatial data and needless expenditure of limited resources.

FGDC Lacks an Up-to-Date and Complete Strategic Plan to Coordinate Its Activities

OMB requires FGDC to prepare and maintain a strategic plan for the development and implementation of the NSDI.[34] Foundational elements of strategic planning, as recognized by federal legislation[35] and OMB guidance,[36] include, among other things, (1) a vision statement; (2) outcome-oriented goals and objectives; (3) a description of how the goals and objectives are to be achieved—including the resources needed and a description of the working relationships with other agencies; (4) a description of how performance goals contribute to the general goals and objectives of the strategic plan; and (5) the identification of external factors that could significantly affect the achievement of the general goals and objectives. Such a plan could help to facilitate coordination among the many geospatial activities that are underway within the federal government and with other stakeholders, and provide a mechanism to measure progress in coordinating geospatial activities and reducing duplication.

FGDC has prepared a strategic plan; however, it is missing key components and has not been kept up-to-date. FGDC's current strategic plan was issued in 2004 and includes (1) a vision statement, (2) three outcome-oriented goals and 13 objectives to be accomplished between 2005 and 2008, and (3) a high-level description of how all but 1 of the 13 objectives are to be achieved. However, its high-level description of the objectives does not include a description of the resources needed to achieve the goals and objectives, or explicitly how the FGDC agencies are to work together to achieve the goals. In addition, the plan does not identify performance measures (such as the percent of the nation for which a given type and standard of data is available) for 9 of the 13 objectives and it does not describe external factors that could affect the achievement of the general goals and objectives, such as the risk of theme-lead agencies not meeting their NSDI development milestones, or limited funding for geospatial investments.

Moreover, the plan does not reflect significant initiatives that the FGDC Steering Committee has engaged in—such as the Geospatial Platform, which did not exist in 2004—and the time frames for FGDC's goals are outdated. For example, the latest time frame associated with the goals in the plan is the year 2007. According to the FGDC Office of the Secretariat Executive Director, the plan needs to be updated; however, he could not provide a time frame, on behalf of the FGDC agencies, for doing so.

The FGDC Office of the Secretariat Executive Director stated that the FGDC has created other strategic planning documents, such as a technical architecture document from 2006 and, more recently, Geospatial Platform planning documents from 2011 and 2012. Individually, these documents contain several foundational elements of strategic planning (e.g., a vision statement, goals and objectives, and discussion of external risks). However, cumulatively they do not represent a comprehensive strategic plan that addresses all aspects of the NSDI for the same unified future time frame.

The Executive Director attributed the lack of a comprehensive strategic plan to competing priorities set by OMB and the FGDC's Steering Committee. Until a comprehensive strategic plan, with meaningful and measurable performance goals, is in place to guide the federal government's geospatial efforts, it is more difficult to achieve the NSDI vision and to hold the FGDC and federal agencies accountable for its development.

Departments and Theme-lead Agencies Have Not Fully Implemented Important Activities for Coordinating and Managing Geospatial Data

OMB has issued guidance, which was followed by an executive order, to federal departments and agencies for effectively coordinating and managing geospatial data to help ensure that they wisely use federal resources in developing the NSDI.[37] According to OMB guidance and the executive order, federal departments and agencies that handle geospatial data are to:

- designate a senior agency official for geospatial information that has departmentwide responsibility, accountability, and authority for geospatial information issues;
- prepare, maintain, publish, and implement a strategy for advancing geographic information and related geospatial data activities appropriate to their mission, and in support of the NSDI strategy;
- develop a policy that requires them to make their geospatial metadata available on the clearinghouse;
- make all metadata associated with geospatial data available on the clearinghouse, and use the metadata standard; and
- adopt internal procedures to ensure that they access the NSDI clearinghouse before they expend funds to collect or produce new geospatial data to determine (1) whether the information has already

been collected by others, or (2) whether cooperative efforts to obtain the data are possible.

As shown in table 1, none of the three federal departments in our review have fully implemented the important activities needed for effectively coordinating and managing geospatial activities within their respective departments.

Table 1. Status of Federal Departments' Implementation of Geospatial Activities

Activity	Commerce	Interior	Transportation
Designate a senior official with departmentwide responsibility	◐	●	◐
Prepare and implement a strategy	○	○	○
Develop a policy for metadata	◐	○	○
Make metadata available on clearinghouse	●	●	●
Adopt procedures for accessing the clearinghouse	○	○	○

Source: GAO analysis of department documentation.
Key:

●= Fully met—the department provided evidence that addressed the criteria.

◐= Partially met—the department provided evidence that addressed about half or a large portion of the criteria.

○= Not met—the department did not provide evidence that addressed the criteria or provided evidence that minimally addressed the criteria.

- *Designate a senior official with departmentwide responsibility.* Only one department—Interior—has designated a senior official with departmentwide responsibility, accountability, and authority. Specifically, in a memo issued in August 2012, the Assistant Secretary for Policy, Management and Budget designated a senior official with departmentwide responsibility, accountability, and authority for geospatial information investments, and for overseeing, coordinating, and facilitating implementation of the department's geospatial-related requirements, policies, activities, and issues.

According to NOAA's Chief Information Officer, he has been designated as Commerce's senior official for geospatial information, but acknowledged that he does not have responsibility and authority for other Commerce geospatial investments, such as those of the Census Bureau. Finally, Transportation has designated a senior official for geospatial information, but this individual does not have departmentwide responsibility, accountability, or authority for geospatial information, as she does not have any insight into, or control over, geospatial activities conducted by the Federal Aviation Administration, one of Transportation's major agencies.

- *Prepare and implement a strategy.* None of the departments have prepared and implemented a strategy for advancing geographic information and related geospatial data activities appropriate to their mission. According to Interior's Deputy Assistant Secretary for Technology, Information, and Business Services, the Geospatial Modernization Blueprint Recommendations and Architectures from 2007 is the department's internal geospatial strategy;38 however, it has not been approved or implemented.
- *Develop a policy for metadata.* None of the departments in our review have established a departmentwide clearinghouse metadata policy. In lieu of a departmentwide policy, two of Commerce's agencies, NOAA and the Census Bureau, have developed policies. Specifically, NOAA's Data Documentation Procedural Directive requires metadata for NOAA environmental data, information, and services to be published to certain national and international clearinghouse portals, and references data.gov. 39 The Census Bureau's metadata policy also mentions the importance of posting metadata to the clearinghouse.[40]
- *Make metadata available on the clearinghouse.* All three departments have made their metadata available on the clearinghouse.

  Most metadata records in the clearinghouse are owned by three federal agencies. Specifically, our analysis of the 441,343 federal records in the centralized geospatial metadata repository, as of February 2012,41 showed that over 99 percent of these records were populated by three agencies: Census Bureau, USGS, and NOAA. See table 2 for the total number of geospatial metadata records by agency.

Most metadata records in the clearinghouse are owned by three federal agencies. Specifically, our analysis of the 441,343 federal records in the centralized geospatial metadata repository, as of February 2012,41 showed that over 99 percent of these records were populated by three agencies: Census Bureau, USGS, and NOAA. See table 2 for the total number of geospatial metadata records by agency.

Table 2. Total Number of Federal Geospatial Metadata Records by Agency

Agency	Number of records	Percentage of total records
Census Bureau	283,855	64.3%
USGS	132,792	30.1
NOAA	23,904	5.4
Other	792	0.2
Total	441,343	100

Source: GAO analysis of General Services Administration data.

Our analysis of the 29 mandatory fields[42] in the 441,343 federal metadata records shows that the metadata records from the three agencies with the majority of geospatial metadata records are largely complete, as shown in table 3. Specifically, nearly all of Census Bureau records had between 24 and 27 mandatory fields completed. In addition, over 95 percent of USGS's and NOAA's records had 28 or 29 of the mandatory fields completed. Finally, the small number of records associated with all other agencies had between 24 and 29 of the mandatory fields completed for virtually all of the records.

Table 3. Percentage of Metadata Records with Range of Mandatory Fields Completed By Agency

Range of mandatory fields	All agencies	Census Bureau	USGS	NOAA	Other agencies
0 – 18	0%	0%	0%	0.3%	0.1%
19 – 23	0.9	0[a]	2.9	0	0.1
24 – 27	64.9	100	1.2	2.0	49.2
28 – 29	34.2	0	95.9	97.7	50.6
Total	100	100	100	100	100

Source: GAO analysis of General Services Administration data.

[a] Our table is rounded to the nearest tenth of a percent. We found that of the 283,855 Census Bureau records, only 50 records, approximately .018 percent, had fewer than 24 of the mandatory fields completed.

- *Adopt procedures for accessing the clearinghouse.* None of the departments have established procedures for searching the clearinghouse before expending funds to acquire or produce geospatial data. While Interior does not have procedures for accessing the clearinghouse, the department's Deputy Assistant Secretary for Technology, Information, and Business Services said that it is commonly practiced and noted that the department usually confers internally with its committees on elevation and orthoimagery[43] prior to acquiring geospatial data.

Theme-lead Agencies Are Not Effectively Managing Data Themes

Department officials attribute the lack of progress in implementing important coordination and management activities to a lack of priority, competing department resources, and in some cases, a lack of awareness. Until the departments implement these activities, they risk acquiring potentially duplicative and costly geospatial data, resulting in the inefficient use of already limited resources.

According to OMB,[44] in order to effectively manage geospatial data and provide the leadership necessary to ensure the national coverage and stewardship of specific geospatial data themes, NSDI-designated theme-lead agencies are to:

- designate a point of contact who is responsible for the development, maintenance, coordination, and dissemination of data using the clearinghouse;
- prepare goals relating to the theme that support the NSDI strategy, and as needed, collect and analyze information from user needs and include those needs in the theme-related goals;
- develop and implement a plan for the nationwide population of the data theme that includes (1) the development of partnership programs with states, tribes, academia, the private sector, other federal agencies, and localities that meet the needs of users; (2) human and financial resource needs; (3) standards, metadata, and the clearinghouse needs; and (4) a timetable for the development for the theme; and
- create a plan to develop and implement theme standards.

As shown in table 4, the three theme-lead agencies in our review have implemented some but not all important geospatial activities.

Table 4. Status of Theme-lead Agencies' Implementation of Geospatial Activities

Activity	NOAA	USGS	BTS
Designate a theme point of contact	●	●	●
Prepare goals and analyze user needs	●	◐	◐
Develop a plan for theme population	●	◐	◐
Develop a standards plan	○	○	○

Source: GAO analysis of agency documentation.

Key

●= Fully met—the agency provided evidence that addressed the criteria.

◐ =Partially met—the agency provided evidence that addressed about half or a large portion of the criteria.

○= Not met—the agency did not provide evidence that addressed the criteria or provided evidence that minimally addressed the criteria.

- *Designate a theme point of contact.* All three agencies have designated a point of contact.
- *Prepare goals and analyze user needs.* One agency has developed goals that recognize and consider user needs for all key datasets in its theme and two agencies have developed goals based on user needs for the major datasets that comprise their themes. Specifically, NOAA's *National Geodetic Survey Ten-Year Plan* includes goals for furthering all of the key datasets in the geodetic control theme.[45] For example, the plan highlights the goals to modernize geometric (horizontal) datum[46] and to modernize the geopotential (vertical) datum,[47] and identifies user needs and why they are important. NOAA has also developed other short-term goals, such as continue education, outreach, development of transition tools and applications, and capacity-building activities to prepare users for the transition to new geometric and geopotential datum; and identify common objectives and find opportunities for cooperative projects and tasks related to standardization and updates to vertical datum. USGS has developed goals for the maintenance of the National Hydrography Dataset and for the Watershed Boundaries Dataset, such as ensuring that datasets

continue to meet user needs; however, these two datasets do not include all datasets within the hydrography theme. Similarly, BTS has developed goals for the development of a comprehensive road centerline[48] dataset; however, the goals do not address all other modes of transportation covered by the data theme, such as railroads, waterways, and virtual airways. The goals recognize differences in user needs for road centerline data, such as basic geometry and naming, support for addressing, and enhanced cartographic displays.

- *Develop a plan for theme population.* One agency has developed a plan that addresses all of the key elements for developing a nationwide plan; one has taken actions consistent with most of the key elements for some of the datasets associated with the theme, but has not integrated all the activities into a single plan that covers the entire theme; and one has developed a plan that addresses some of the key elements, but for only one of the major datasets in the theme. NOAA has developed a plan, its National Geodetic Survey Ten-Year Plan, which provides a strategy for how NOAA intends to modernize and populate its data theme. The plan includes the development of partnership programs with states, academia, federal agencies, and other stakeholders; identifies the need to address human and financial resource needs; identifies needs for standards, metadata, and the clearinghouse; and advances a timetable for the development of the theme. In contrast, USGS does not have a plan for the population of the hydrography theme; however, it has taken actions to (1) develop partnerships; (2) address human and financial resources; and (3) identify needs for standards, metadata, and the clearinghouse for the National Hydrography Dataset, which is one dataset within the hydrography theme. BTS has developed a strategic plan for the development of a nationwide road centerline dataset. However, this plan does not address all other modes of transportation covered by the data theme, such as railroads, waterways, and virtual airways. These officials stated that a strategic plan for connecting all of the transportation datasets in an intermodal manner is needed; however, there are no plans to create such a plan in the immediate future.
- Develop a standards plan. None of the agencies have developed a plan to develop and implement standards; while NOAA recognized the need for a plan, both USGS and BTS stated that such a plan was not needed for their themes—USGS because of the maturity of their existing datasets, and BTS because they collect data from states and

counties in various formats, and then standardize the data themselves. However, without a plan to maintain existing standards or anticipate new standards, these agencies risk potential future difficulties exchanging and sharing geospatial datasets.

Theme-lead agency officials attribute the lack of progress in implementing these activities to competing priorities, limited resources, and the perceived lack of need for some plans. Until agencies implement these activities, they will be challenged to effectively manage important geospatial activities, wisely use limited resources, and risk engaging in potential duplicative geospatial acquisition efforts.

OMB Does Not Have Complete and Reliable Information to Identify Duplicative Geospatial Investments

OMB has oversight responsibilities for federal IT systems and acquisition activities—including GIS—to help ensure their efficient and effective use. According to OMB Office of E-Government staff members, OMB relies primarily on the annual budget process to identify potentially duplicative geospatial investments; specifically, the exhibit 53s and 300s.

However, OMB's Office of E-Government staff members acknowledged that these two sources may not in all cases provide the necessary information to allow OMB to identify potentially duplicative investments or accurately quantify the amount of federal dollars spent on geospatial datasets for three primary reasons. First, these staff members stated that some federal agencies may not classify investments in geospatial data as "information technology" (such as satellites), meaning that they would not be captured in exhibit 53s. OMB staff members stated that agencies are to determine what qualifies as an IT investment and stated that there are variations in the way that agencies interpret the definition of IT. Second, agencies do not always appropriately classify geospatial investments as "geospatial services" using the Federal Enterprise Architecture codes. Our analysis of the fiscal year 2013 exhibit 53s for the three departments that we reviewed showed that only 5 of their 24 key datasets[49]—1 of NOAA's 6 geodetic control datasets,[50] and 4 of USGS's 7 hydrography datasets[51]—were included in the departments' exhibit 53s. Further, only 1 of these investments[52] was identified with the geospatial services code, as required by OMB's fiscal year 2013 budget formulation guidance.[53] Third, given that the geospatial data may be only one component of an IT investment or capital asset, even if it were included in the agencies' exhibit 53s or 300s, OMB would have difficulties in identifying the geospatial component, and the associated dollars, without having a detailed discussion with individuals responsible for each investment.

OMB staff members stated that they do not have a complete picture of how much money is being spent on geospatial investments across the federal government because, as noted above, what is being reported may not capture all geospatial spending, and the data have not been reliable. In 2006 and 2007, OMB made two data calls directly to federal agencies to determine federal agencies' spending on geospatial investments. However, according to OMB, neither of these data calls provided the agency with complete and reliable information—largely because agencies either provided incomplete information, or did not respond at all. Although the data may not be complete, those agencies that did respond reported that they planned to spend about $1.89 billion in geospatial data and services between fiscal years 2007 and 2009, of which about $1.53 billion, or about 81 percent, was to be on geospatial data. OMB staff members stated that OMB has not made any additional data collections outside of the IT reporting process (i.e., exhibit 53s and 300s) for geospatial data investments since 2007, and as of September 2012, has no current plans for any new collections.

OMB staff members stated that, although eliminating duplication in geospatial investments is important, OMB's recent efforts have focused on other commodity IT areas with higher spending and cyber security ramifications, such as reducing the numbers of federal data centers and internet connections maintained by the government. However, without complete and reliable information on the federal government's investments in geospatial data, including the amount of federal dollars spent, OMB does not have the necessary information to make a fact-based decision about the potential priority of geospatial information in relation to other activities.

Federal Agencies Have Coordinated Specific Investments, but Duplicative Geospatial Data Exists

According to FGDC Secretariat officials, departments and agencies have taken steps to coordinate geospatial data related to their respective data themes.

- Interior participates in several efforts aimed at coordinating one of its themes, orthoimagery. For example, in conjunction with the U.S. Department of Agriculture, Interior participates in the National Agricultural Imagery Program, which was developed to obtain one-meter resolution, "leaf-on" imagery[54] for the 48 continental states on a 3-year cycle. In addition, Interior, in conjunction with the Department of Defense, participates in the Urban Area Imagery Program. The goal of this program is to acquire one-foot resolution, "leaf-off" imagery[55] for 133 of the nation's largest or most important urban areas (such as state capitals) on a 2-to-4-year cycle. According to FGDC Secretariat officials, for both of these programs, members acquire the imagery, assure the quality of the imagery, and distribute it to the participants in the program. Each participant in these programs benefits from the imagery acquired, as do all other federal, state, and local agencies,
- the private sector (such as companies that provide mapping services on the Internet), nonprofit organizations, and members of the public interested in this type of imagery.
- Interior also participates in a program for another theme, elevation. Specifically, in conjunction with other departments, Interior participates in the National Digital Elevation Program, which is intended to acquire detailed elevation data using advanced technologies to support various programmatic needs, such as USGS's Coastal and Marine Geology program, the U.S. Department of Agriculture's Natural Resources Conservation Service (which supports precision farming), and the Department of Homeland Security's Federal Emergency Management Agency's Risk Mapping, Assessment, and Planning program. The participating agencies seek other federal agency and state and local government participation in elevation acquisition projects as part of their program design. According to Secretariat officials, agencies generally coordinate their plans and funding, and execute the projects independently. In addition, the data acquired by projects are available to others.
- USGS further coordinated the elevation theme when, in conjunction with its partners, it sponsored the National Enhanced Elevation Assessment. According to USGS officials, the assessment was conducted to investigate the potential to obtain greater benefits and efficiencies from the growing interest in using data from advanced technologies, and concluded that moving to a coordinated national program has the potential to produce new benefits in excess of $1

billion annually. USGS and its federal partners intend to use the recommendations from the assessment to improve on the current coordination-based approach among federal agencies and nonfederal organizations.

- As the theme-lead agency of the hydrography theme, USGS has led efforts to coordinate the population of the hydrography theme. According to USGS officials, it created a partnership for the development of two separate datasets associated with hydrography: the National Hydrography Dataset and the Watershed Boundaries Dataset. These datasets were created by pooling funding and resources from several federal, state, and local agencies into a single effort. This partnership supports the missions of several federal agencies: USGS and the U.S. Forest Service for mapping and analysis projects, the Environmental Protection Agency and the Department of Homeland Security for analysis projects, and the Census Bureau for mapping. Numerous state agencies also participate in order to meet reporting requirements of the Federal Water Pollution Control Act. This partnership also has data stewardship agreements in place between USGS and 35 states. This data stewardship activity is based on the input of local organizations knowledgeable about hydrography in their immediate area.

While agencies have taken actions to coordinate geospatial data investments, recent reports, as well as officials from state and local associations and the FGDC Advisory Committee have all stated that duplicative geospatial data investments continue across all levels of government. For example, according to Transportation's *Transportation for the Nation Strategic Plan*, dated May 2011, duplication exists in the acquisition of nationwide road centerline data across federal agencies and other levels of government, resulting in millions of wasted taxpayer dollars.[56] The report identified several initiatives that are currently independently acquiring road centerline data:

- Census Bureau's Topologically Integrated Geographic Encoding and Referencing (TIGER) system, which uses data procured from local sources for census enumeration and demographic applications. These data were built and are maintained by the Census Bureau.

- USGS's National Map website, which uses licensed data from a commercial provider to create viewable maps on the National Map. These data are managed by USGS.[57]
- The Department of Defense's Homeland Security Infrastructure Program, which uses licensed commercial data procured by the National Geospatial-Intelligence Agency for emergency management.

In addition, a subcommittee of the National Geospatial Advisory Committee has, at the request of the FGDC, been evaluating the need for a national address database, assessing potential concerns with such a database, and identifying possible approaches for its development. According to a National Geospatial Advisory Committee official, several federal agencies collect, purchase, or lease address information in a noncoordinated fashion. This subcommittee is in the process of finalizing a report[58] for the full committee that would assess, among other things, the benefits, potential savings, and efficiencies that could be realized from the development of a national address database.

Further, in a report on land parcel data, the National Academy of Sciences [59] stated that the lack of nationally integrated land parcel data has led to duplication of effort among various levels of government and between the public and private sector, such as the Department of Housing and Urban Development, the U. S. Forest Service, insurance companies, and private companies that list home values and sell parcel maps.[60] In addition, a National Geospatial Advisory Committee representative stated that a commercial provider leases the same proprietary parcel data to six federal agencies: the Department of Housing and Urban Development, the Department of Homeland Security, the Federal Bureau of Investigation, the Small Business Administration, the Federal Deposit Insurance Corporation, and the Federal Reserve. In recent reports, the Congressional Research Service found that a coordinated approach to federally managed parcel data still did not exist and that the best method for obtaining an accurate tally of federal lands is to contact each land management agency directly.[61]

Representatives from the National States Geographic Information Council[62] stated that federal agencies are investing in geospatial data that exist at the state and local level, noting that duplicative data continue to be procured in such areas as imagery, elevation, road centerlines, and address points.

Conclusions

The long-standing problem of effectively coordinating federal geospatial investments to reduce redundancies has yet to be resolved. In particular, the FGDC has not established a framework for governmentwide management of themes and datasets; provided geospatial information users with the means to identify planned data acquisitions; and developed and maintained a national strategy to guide the development of the NSDI, with associated metrics to measure progress and ensure accountability. Similarly, federal departments and agencies have not yet implemented long-standing OMB guidance intended to ensure the efficient use of limited federal resources and the effective stewardship of geospatial data themes, including developing and implementing a strategy for advancing geospatial activities related to their mission; and implementing policies, procedures, and plans for effectively coordinating and managing geospatial data, standards, and the clearinghouse. Moreover, OMB has not established an effective mechanism to identify the amount of the federal budget being spent on geospatial investments as well as potentially duplicative geospatial investments.

The FGDC, federal agencies, and OMB have each indicated that the lack of progress in implementing these important coordination activities is because they have been focusing on other priorities. However, while the extent of duplication in geospatial investments is unknown, it is estimated that billions of dollars are being spent across the federal government on geospatial investments. Further, many mission-critical applications, such as those used to respond to natural disasters—floods, hurricanes, and fires—depend on geospatial information to protect lives and property. Thus, it is important that the data acquired to support these critical functions be done in a timely and coordinated manner, with minimal duplication.

Until a comprehensive national strategy is in place, the FGDC develops and implements guidance and tools to effectively coordinate governmentwide geospatial activities, and federal agencies establish and implement the policies, procedures, and plans to coordinate their geospatial activities, the vision of the NSDI to improve the coordination and use of geospatial information will likely not be fully realized and duplicative investments will likely continue. Further, until OMB establishes a way to obtain reliable information about federal geospatial investments, OMB will not be able to identify potentially duplicative geospatial investments. Unless the FGDC, federal departments and agencies, and OMB decide that investments in geospatial information are a priority, these investments will remain uncoordinated, and although the extent of duplication is unknown, the federal government will continue to acquire duplicative geospatial information and waste taxpayer dollars.

RECOMMENDATIONS FOR EXECUTIVE ACTION

To better facilitate the coordination of—and accountability for—the estimated billions of dollars in federal geospatial investments, and to reduce duplication, we recommend that the Secretary of the Interior, as the FGDC Chair, direct the FGDC Steering Committee to take the following three actions.

- Establish a time frame for completing a plan to facilitate the implementation of OMB's portfolio management guidance, and develop and implement the plan within the established time frame. The plan, at a minimum, should include goals and performance measures, and the FGDC should report annually to OMB on the progress made on efforts to improve coordination and reduce duplication among themes.
- Develop and implement guidance for identifying planned geospatial investments using the Geospatial Platform, and establish a time frame for doing so.
- Establish a time frame for creating and updating a strategic plan to improve coordination and reduce duplication, and create and implement the plan within the established time frame. The plan, at a minimum, should include (1) a vision statement for the NSDI; (2) outcome-oriented goals and objectives that address all aspects of the NSDI; (3) a description of how the goals and objectives are to be achieved, including a description of the resources needed to achieve the goals and objectives and how the FGDC is to work with other agencies to achieve them; (4) performance measures for achieving the stated goals; and (5) external factors that could affect the achievement of the goals and objectives.

To help ensure the success of department's efforts to improve geospatial coordination and reduce duplication, we recommend that the Secretary of Commerce designate a senior agency official who has departmentwide responsibility, accountability, and authority for geospatial information issues.

We further recommend that the Secretary of Commerce direct the designated senior official for geospatial information to take the following three actions.

- Prepare, maintain, publish, and implement a strategy for advancing geographic information and related geospatial data activities appropriate to its mission.
- Develop a policy that requires the department to make its geospatial metadata available on the clearinghouse.
- Develop and implement internal procedures to ensure that it accesses the NSDI clearinghouse before it expends funds to collect or produce new geospatial data to determine (1) whether the information has already been collected by others and (2) whether cooperative efforts to obtain the data are possible.

Further, to improve the department's management of its geodetic control theme, we recommend that the Secretary of Commerce direct the geodetic control theme point of contact to create and implement a plan to develop and implement geodetic control theme standards.

We recommend that the Secretary of the Interior direct the designated senior official for geospatial information to take the following three actions.

- Prepare, maintain, publish, and implement a strategy for advancing geographic information and related geospatial data activities appropriate to its mission.
- Develop a policy that requires the department to make its geospatial metadata available on the clearinghouse.
- Develop and implement internal procedures to ensure that it accesses the NSDI clearinghouse before it expends funds to collect or produce new geospatial data to determine (1) whether the information has already been collected by others and (2) whether cooperative efforts to obtain the data are possible.

We further recommend that the Secretary of the Interior direct the hydrography theme point of contact to take the following three actions.

- Prepare goals relating to all datasets within the hydrography theme that support the NSDI, and as needed, collect and analyze information from user needs and include those needs in the theme-related goals.

- Develop and implement a plan for the nationwide population of the hydrography theme that addresses all datasets within the theme; and that includes (1) the development of partnership programs with states, tribes, academia, the private sector, other federal agencies, and localities that meet the needs of users; (2) human and financial resource needs; (3) standards, metadata, and the clearinghouse needs; and (4) a timetable for the development for the theme.
- Create and implement a plan to develop and implement hydrography theme standards.

We recommend that the Secretary of Transportation designate a senior agency official who has departmentwide responsibility, accountability, and authority for geospatial information issues. We further recommend that the Secretary of Transportation direct the designated senior official for geospatial information to take the following three actions.

- Prepare, maintain, publish, and implement a strategy for advancing geographic information and related geospatial data activities appropriate to its mission.
- Develop a policy that requires the department to make its geospatial metadata available on the clearinghouse.
- Develop and implement internal procedures to ensure that it accesses the NSDI clearinghouse before it expends funds to collect or produce new geospatial data to determine (1) whether the information has already been collected by others and (2) whether cooperative efforts to obtain the data are possible.

We further recommend that the Secretary of Transportation direct the transportation theme point of contact to take the following three actions.

- Prepare goals relating to all datasets within the transportation theme that support the NSDI, and as needed, collect and analyze information from user needs and include those needs in the theme-related goals.
- Develop and implement a plan for the nationwide population of the transportation theme that addresses all datasets within the theme; and that includes (1) the development of partnership programs with states, tribes, academia, the private sector, other federal agencies, and localities that meet the needs of users; (2) human and financial resource needs;

(3) standards, metadata, and the clearinghouse needs; and (4) a timetable for the development for the theme.

- Create and implement a plan to develop and implement transportation theme standards.

Further, to improve OMB oversight of geospatial information and assets, and minimize duplication of federal geospatial investments, we recommend that the Director of OMB develop a mechanism, or modify existing mechanisms, to identify and report annually on all geospatialrelated investments, including dollars invested and the nature of the investment.

AGENCY COMMENTS AND OUR EVALUATION

We received written, e-mail, or oral responses on a draft of this report from Interior, Commerce, Transportation, as well as OMB and the General Services Administration. These responses are summarized below.

In written comments, signed by the Assistant Secretary for Policy, Management and Budget, and reprinted in appendix III, Interior generally agreed with our recommendations. The department stated that it recognizes the need to more fully implement the portfolio management requirements described in the OMB Circular A-16 supplemental guidance and is already actively working to develop tools that will help agencies identify planned geospatial investments in the Geospatial Platform. With respect to our recommendations aimed at improving Interior's management of its geospatial investments, the department stated that it is beginning to take actions to implement our recommendations, including developing an Geospatial Advisory Committee for the department—which is intended to provide leadership and direction for the development of a comprehensive geospatial technical strategy for the department—as well as developing procedures for making metadata available on the clearinghouse. The department further stated that it is committed to working with OMB and its partner agencies to address our recommendations in a timely manner.

In addition, the department stated that its efforts to lead and coordinate activities of the FGDC and the Geospatial Platform continue to accrue great benefits to the federal community and U.S. citizens. In particular, Interior stated that through these efforts, it will be able to implement the supplemental guidance to OMB Circular A-16, and realize the tremendous potential of the Geospatial Platform. We support the department's efforts as evidenced by our recommendations to the department aimed at furthering agencies' implementation of the supplemental guidance and use of the Geospatial Platform to identify planned geospatial investments. Interior also provided technical comments, which we have incorporated in the report as appropriate.

In written comments, signed by the Acting Secretary of Commerce, and reprinted in appendix IV, Commerce stated that the department and NOAA agree with our recommendations, and described actions planned to implement them. Commerce also stated that it appreciates the work that we have done to improve coordination in managing geospatial investments.

In oral comments, Transportation officials, including the department's Geospatial Information Officer/BTS Director of Geospatial Information Systems, neither agreed nor disagreed with our recommendations, and provided two comments on our draft report. First, Transportation officials stated that, as of October 2012, the department's metadata files are available on the geoplatform.gov and the geo.data.gov sites, and provided supporting evidence. As a result, we revised our report to acknowledge that the department has made its metadata associated with geospatial data available on the clearinghouse, and removed the recommendation that it does so. Second, Transportation officials stated that they believed that the department should have received partial credit for having prepared, maintained, published, and implemented a strategy for advancing geographic information and related geospatial data activities appropriate to the department's mission, and in support of the NSDI strategy. Specifically, Transportation officials stated that the department's *Transportation for the Nation Strategic Plan* partially satisfies these criteria because it includes a strategy for collecting and maintaining road centerline data, which represents the vast majority of travel in terms of both passengers and freight. We agree that transportation by road is a major component of the transportation data theme. In fact, our assessment of this plan was the basis for the partial rating for two of the criteria related to BTS's (the theme-lead agency) management of the transportation data theme: "prepared goals and analyzed user needs," and "developed a plan for theme population." However, the strategic plan does not include a strategy for advancing all the department's geographic information and related geospatial data activities, nor does it describe how the department and its agencies are to coordinate their geospatial efforts to support the department's mission. In particular, the plan does not address geospatial themes other than transportation—including elevation and imagery—in which Federal Aviation Administration officials stated that their agency also makes investments. Therefore, we believe that the department's *Transportation for the Nation*

Strategic Plan does not constitute a departmentwide geospatial plan. In addition, Transportation officials provided a technical comment, which we incorporated into the draft.

In comments provided via e-mail, a paralegal specialist in OMB's Office of General Counsel, on behalf of OMB, stated that OMB concurs with the need for improved collection of geospatial-related investments, but believes that it should only be achieved through improvements to broader reporting mechanisms for IT investments and data assets, and not by developing new and separate mechanisms specifically for geospatialrelated assets. OMB further noted that it would be helpful if we clarified our recommendation to acknowledge that a new process is not required or expected. The decision as to whether OMB should develop a new mechanism, or improve an existing mechanism, should be made based on whichever option will be most successful in collecting the necessary information. We modified our recommendation to reflect this. OMB also provided technical comments, which we incorporated as appropriate.

In e-mail comments provided by a Management and Program Analyst in the GAO/IG (Inspector General) Response Audit Division, the General Services Administration stated that it had no formal comments on the draft report.

David A. Powner
Director
Information Technology Management Issues

APPENDIX I: OBJECTIVE, SCOPE, AND METHODOLOGY

Our objective was to determine the extent to which the federal government has established and effectively implemented policies and procedures for coordinating its investments in geospatial data and avoiding duplication. To address this objective, we focused on governmentwide activities to implement the National Spatial Data Infrastructure (NSDI), including efforts of the Federal Geographic Data Committee (FGDC), as well as those within selected departments.

To evaluate federal departments' efforts to implement the NSDI, we first identified the nine framework themes,[1] as identified in Office of Management and Budget (OMB) Circular A-16. These framework themes are cadastral, cadastral (offshore), digital orthoimagery, elevation bathymetric, elevation terrestrial, geodetic control, government units, hydrography, and transportation; and are described in appendix II. From those nine themes, we then randomly selected three themes and identified the federal departments and agencies responsible for managing the themes. The three departments, theme-lead agencies, and selected themes are:

- Department of Commerce (Commerce)—National Oceanic and Atmospheric Administration (NOAA)—Geodetic Control;
- Department of the Interior (Interior)—U. S. Geological Survey (USGS)—Hydrography; and
- Department of Transportation (Transportation)—Bureau of Transportation Statistics (BTS)—Transportation.

We then reviewed FGDC and federal department documentation, such as policies, procedures, strategic plans, implementation plans, technical documentation of standards and metadata, committee charters and meeting minutes, and budget documentation. We assessed this information against responsibilities identified in Executive Order 12906,[2] OMB Circular A-16,[3] OMB M-06-07,[4] and OMB M-11-03;[5] identified any discrepancies, and discussed them with the relevant agency officials.

To determine the completeness of the federal agency metadata records in the clearinghouse and to determine which agencies contributed records to geo.data.gov, we obtained an extract of the contents of geo.data.gov as of February 22, 2012. We then recreated a database of the metadata records. We reviewed FGDC guidance [6] to identify what are considered to be the mandatory data elements, which total 29. For each record in the database, we determined if the metadata value for each of the 29 mandatory data fields had information or was blank. In addition, we used the metadata records to determine what agencies contributed records by examining the data fields that indicated the origin and publisher of the data.

In order to assess the reliability of the clearinghouse data that we analyzed, we reviewed FGDC documentation, the General Services Administration's written responses to questions, and interviewed officials familiar with the clearinghouse data in order to gain an understanding of the controls around the creation and maintenance of the clearinghouse data. We determined that our recreation of the database had no material effect on our analysis and that the database, as recreated, was sufficiently reliable for our purposes.

To determine whether OMB had complete and reliable information to identify duplicative geospatial investments, we reviewed OMB's most recent data calls for geospatial data from 2006 and 2007, OMB Circular A-11,[7] and department budget submissions. We compared the list of key datasets for the themes in our sample to budget documentation available in OMB's fiscal year 2013 exhibit 53 to determine the extent to which the agencies identified these datasets as investments in information technology (IT) and geospatial services. To determine the reliability of the data on the IT Dashboard, we reviewed recent GAO reports that identified issues with the accuracy and reliability of agency data on the IT Dashboard. We determined that the data were sufficiently reliable for the purpose of this report, which is to determine the extent to which departments' key datasets were included as IT investments and coded as geospatial services in the departments' respective exhibit 53s for fiscal year 2013.

To identify potential duplicative geospatial investments, we reviewed recent reports[8] from the FGDC, Transportation, the Congressional Research Service, the National Geospatial Advisory Committee,[9] and the National Academy of Sciences,[10] and spoke with FGDC and the National States Geographic Information Council officials.[11]

We also interviewed FGDC officials, including the Chair and Vice-Chair of the FGDC; the Executive Director of the FGDC Office of the Secretariat; the representative for the Managing Partner of the Geospatial Platform; OMB Office of E-Government staff members; department and agency officials responsible for coordinating geospatial investments within their respective agencies as well as theme-lead points of contact within those agencies; General Services Administration officials responsible for managing the geospatial clearinghouse; the Chair of the National Geospatial Advisory Committee; and the President and Washington Liaison of the National States Geographic Information Council.

We conducted this performance audit from November 2011 to November 2012 in accordance with generally accepted government auditing standards. Those standards require that we plan and perform the audit to obtain sufficient, appropriate evidence to provide a reasonable basis for our findings and conclusions based on our audit objective. We believe that the evidence obtained provides a reasonable basis for our findings and conclusions based on our audit objective.

APPENDIX II: COMPARISON BETWEEN PROPOSED THEMES AND EXISTING THEMES

FGDC proposed theme and description	Corresponding A-16 existing theme and description
Biota—Biota pertain to, or describe the dynamic processes, interactions, distributions, and relationships between and among organisms and their environments.	*Biological Resources*—This dataset includes data pertaining to or descriptive of (nonhuman) biological resources and their distributions and habitats, including data at the suborganismal (genetics, physiology, anatomy, etc.), organismal (subspecies, species, systematics), and ecological (populations, communities, ecosystems, biomes, etc.) levels. *Vegetation*—Vegetation data describe a collection of plants or plant communities with distinguishable characteristics that occupy an area of interest.
Cadastre—This theme describes past, current, and future rights and interests in real property, including the spatial information necessary to describe geographic extents. Rights and interests are benefits or enjoyment in real property that can be conveyed, transferred, or otherwise allocated to another for economic remuneration. Rights and interests are recorded in land record documents. The spatial information necessary to describe geographic extents includes surveys and legal description frameworks, such as the Public Land Survey System, as well as parcel-by-parcel surveys and descriptions. This theme does not include federal government or military facilities.	*Cadastral*[a] —Cadastral data describe the geographic extent of past, current, and future rights, title, and interests in real property, and the framework to support the description of that geographic extent. The geographic extent includes survey and description frameworks, such as the Public Land Survey System, as well as parcelby-parcel surveys and descriptions. *Cadastral (Offshore)*[a]—Offshore cadastre is the land management system used on the Outer Continental Shelf. It extends from the baseline to the extent of U.S. jurisdiction. Existing coverage is currently limited to the conterminous United States and portions of Alaska. *Public Land Conveyance (patent) Records*—Public land conveyance data are the records that describe all past, current, and future rights, title, and interests in real property. This is a system of storage, retrieval, and dissemination of documents describing the rights, title, and interests of parcels. *Federal Land Ownership Status*—Federal land ownership status includes the establishment and maintenance of a system

FGDC proposed theme and description	Corresponding A-16 existing theme and description
	for the storage and dissemination of information describing all title, estates, or interests of the federal government parcels of real and mineral property. The ownership status system is the portrayal of title for all such federal estates or interests in land.
Climate and Weather—Climate and weather describes meteorological conditions, including temperature, precipitation, and wind, that characteristically prevail in a particular region over a long period of time. Weather is the state of the atmosphere at a given time and place, with respect to variables, such as temperature, moisture, wind velocity, and barometric pressure.	*Climate*—Climate data describe the spatial and temporal characteristics of the Earth's atmosphere, hydrosphere, and land surface system. These data represent both model-generated and observed environmental information, which can be summarized to describe surface, near surface, and atmospheric conditions over a range of scales.
Cultural Resources—This theme describes features and characteristics of a collection of places of significance in history, architecture, engineering, or society. It includes national monuments and icons.	*Cultural Resources*—The cultural resources theme includes historic places, such as districts, sites, buildings, and structures of significance in history, architecture, engineering, or culture. Cultural resources also encompass prehistoric features as well as historic landscapes. *Geographic Names*—This dataset contains data or information on geographic place names deemed official for federal use by the U.S. Board on Geographic Names as pursuant to 80 Cong. Ch. 330. Geographic Names information includes both the official place name (current, historical, and aliases) and direct (i.e., geographic coordinates) and indirect (i.e., state and county where place is located) geospatial identifiers. This information is categorized as populated places, schools, reservoirs, parks, streams, valleys, and ridges.
Elevation—Elevation is the measured vertical position of the Earth surface and other landscape or bathymetric features relative to a reference datum typically related to sea level. These points normally describe bare Earth positions but may also describe the top surface of buildings and other objects, vegetation structure, or	*Elevation Bathymetrica* —The bathymetric data for Inland and Intercoastal waterways is highly accurate bathymetric (i.e., the measurement of water depths) sounding information collected to ensure that federal navigation channels are maintained to their authorized depths. Bathymetric survey activities support the nation's critical

Appendix II. (Continued)

FGDC proposed theme and description	Corresponding A-16 existing theme and description
submerged objects. Elevation data can be stored as a threedimensional array or as a continuous surface, such as a raster, triangulated irregular network, or contours. Elevation data may also be represented in other derivative forms, such as slope, aspect, ridge and drainage lines, and shaded relief.	nautical charting program. These data are also used to create Electronic Navigational Charts. *Elevation Terrestrial*[a] —These data contain georeferenced digital representations of terrestrial surfaces, natural or manmade, that describe vertical position above or below a datum surface
Geodetic Control—This theme includes a collection of control points that provide a common reference system for establishing coordinates for geographic data.	*Geodetic Control*[a] —Geodetic control provides a common reference system for establishing coordinates for all geographic data. All NSDI framework data and users' applications data require geodetic control to accurately register spatial data. The National Spatial Reference System is the fundamental geodetic control for the United States.
Geology—Geology is geographically-referenced data pertaining to the origin, history, composition, structure, features, and processes of the solid Earth, both onshore and offshore. It includes geologic, geophysical, and geochemical maps, stratigraphy, paleontology, geochronology, mineral and energy resources, and natural hazards, such as earthquakes, volcanic eruptions, coastal erosion, and landslides. The theme does not include soils.	*Geologic*—The geologic spatial data theme includes all geologic mapping information and related geoscience spatial data (including associated geophysical, geochemical, geochronologic, and paleontologic data) that can contribute to the National Geologic Map Database as pursuant to Public Law 106-148. *Offshore Minerals*—Offshore minerals include minerals occurring in submerged lands. Examples of marine minerals include oil, gas, sulfur, gold, sand and gravel, and manganese.
Governmental Units—This theme includes data that describe political, governmental, and administrative (management) type boundaries that are used to manage people and resources. It includes geopolitical boundaries (county, parish, state, city, etc.), tribal boundaries, federal land boundaries, federal regions, international boundaries, and governmental administrative units, such as congressional districts, international lines of separation, limits,	*Governmental Units*[a] —These data describe, by a consistent set of rules and semantic definitions, the official boundaries of federal, state, local, and tribal governments as reported/certified to the U.S. Census Bureau by responsible officials of each government for purposes of reporting the nation's official statistics. *International Boundaries*—International boundary data include both textual information to describe, and geographic

FGDC proposed theme and description	Corresponding A-16 existing theme and description
zones, enclaves, exclaves, special areas between states and dependencies, and all jurisdictional offshore limits within U.S. sovereignty. Boundaries associated with natural resources, demography, and cultural entities are excluded and can be found in the appropriate subject themes.	information system (GIS) digital cartographic data to depict, both land and maritime international boundaries, other lines of separation, limits, zones, enclaves, exclaves, and special areas between states and dependencies. *Marine Boundaries*—Marine boundaries depict offshore waters and seabeds over which the United States has sovereignty and jurisdiction.
Imagery—The imagery theme includes georeferenced images of the Earth's surface that have been collected using aerial photography or satellite data. Orthoimagery is prepared through a geometric correction process known as orthorectification to remove image displacements due to relief and sensor characteristics. This process allows for their use as base maps for digital mapping and analyses in a GIS. Specific imagery data sets created through image interpretation and classification, such as a land cover image, can be found under themes specific to the subject matter. This theme includes imagery, such as Landsat, National Agriculture Imagery Program, and Digital Orthophoto Quarter Quadrangle.	*Digital Ortho Imagery*[a] —This dataset contains georeferenced images of the Earth's surface where object displacement has been removed for sensor distortions and orientation, and terrain relief. Digital orthoimages have the geometric characteristics of a map and image qualities of a photograph.
Land Use/Land Cover—This theme refers collectively to natural and man-made surface features that cover the land (Land Cover) and to the primary ways in which land cover is used by humans (Land Use). Examples of Land Cover may be grass, asphalt, trees, bare ground, and water. Examples of Land Use may be urban, agricultural, ranges, and forest areas.	*Earth Cover*—The Earth cover theme uses a hierarchical classification system based on observable form and structure, as opposed to function or use. This system transitions from generalized to more specific and detailed class divisions, and provides a framework within which multiple land cover and land use classification systems can be cross-referenced. This system is applicable everywhere on the surface of the Earth. This theme differs from the Vegetation and Wetlands themes, which provide additional detail.
Real Property—The real property theme includes data that describes the spatial representation (location) of real property	*Buildings and Facilities*—The facility theme includes federal sites or entities with a geospatial location deliberately established

Appendix II. (Continued)

FGDC proposed theme and description	Corresponding A-16 existing theme and description
entities, typically consisting of one or more of the following: unimproved land, a building, a structure, site improvements, and the underlying land. Complex real property entities (i.e., facilities) are used for a broad spectrum of functions or missions. This theme focuses on spatial representation of real property assets only and does not seek to describe special purpose functions of real property, such as those found in the Cultural Resources, Transportation, or Utilities themes.	for designated activities; a facility database might describe a factory, military base, college, hospital, power plant, fishery, national park, office building, space command center, or prison. *Housing*—This theme includes geographic data on homeownership rates, including many attributes such as the Department of Housing and Urban Development's revitalization zones, location of various forms of housing assistance, first-time homebuyers, underserved areas, and race.
Soils—This theme includes data that depict the geography and attributes of the many kinds of soils found in the landscape at both large and small map scales. A living dynamic resource providing a natural medium for plant growth and habitat for living organisms, soil recycles nutrients and wastes, stores carbon, and purifies water supplies. Soil has distinct layers (called "horizons") that, in contrast to underlying geologic material, are altered by the interactions of climate, landscape features, and living organisms over time.	*Soils*—Soil data consist of georeferenced digital map data describing the spatial distribution of the various soils that cover the Earth's surface, and attribute data describing the proportionate extent of the various soils as well as the physical and chemical characteristics of those soils. The physical and chemical properties are based on observed and measured values, as well as model-generated values. This theme also includes model-generated assessments of the suitability or limitations of the soils to various land uses.
Transportation—Transportation data describe means and aids for conveying persons and goods. The transportation system includes both physical and nonphysical components related to all modes of travel that allow the movement of goods and people between locations.	*Transportation*[a] —Transportation data are used to model the geographic locations, interconnectedness, and characteristics of the transportation system within the United States. The transportation system includes both physical and nonphysical components representing all modes of travel that allow the movement of goods and people between locations. *Transportation (Marine)*—The Navigation Channel Framework consists of highly accurate dimensions (geographic coordinates for channel sides, centerlines, wideners, turning basins, and river mile markers) for every federal navigation channel maintained by the U.S. Army Corps of Engineers. The

FGDC proposed theme and description	Corresponding A-16 existing theme and description
	navigation framework will provide the basis for the marine transportation theme of the geospatial data framework.
Utilities—This theme includes the means, aids, and usage of facilities for producing, conveying, distributing, processing, and disposing of public and private commodities, including power, energy, communications, natural gas, and water. It includes subthemes for Energy and Communications.	There are not any corresponding existing themes.
Water—Inland—The Water—Inland theme describes interior hydrologic features and characteristics, including classification, measurements, location, and extent. It includes aquifers, watersheds, wetlands, navigation, water quality, water quantity, and groundwater information.	*Hydrography*[a] —This data theme includes surface water features, such as lakes, ponds, streams and rivers, canals, oceans, and coastlines. *Watershed Boundaries*—This data theme encodes hydrologic, watershed boundaries into topographically defined sets of drainage areas that are organized in a nested hierarchy by size, and based on a standard hydrologic unit coding system. *Flood Hazards/Flood Plain*—This theme includes documentation of the boundaries and elevations of the 1 percent annual chance (100-year) flood for communities across the country. The National Flood Insurance Program has prepared flood hazard data for approximately 18,000 communities. The primary information prepared for these communities is for the 1 percent annual chance (100-year) flood, and includes documentation of the boundaries and elevations of that flood. *Wetlands*—The wetlands data layer provides the classification, location, and extent of wetlands and deepwater habitats.
Water—Oceans and Coasts—This theme includes datasets that describe features and characteristics of salt water bodies (tides, tidal waves, coastal information, reefs) and features and characteristics that represent the intersection of the land with the water surface (shorelines), the lines from which	*Baseline/Maritime Zones*—Baseline represents the line from which maritime zones and limits are measured. Examples of these limits include the territorial sea, contiguous zone, and exclusive economic zone. *Outer Continental Shelf Submerged Lands*—

Appendix II. (Continued)

FGDC proposed theme and description	Corresponding A-16 existing theme and description
the territorial sea and other maritime zones are measured (baseline maritime), and lands covered by water at any stage of the tide (outer continental shelf), as distinguished from tidelands, which are attached to the mainland or an island and cover and uncover with the tide.	These data include lands covered by water at any stage of the tide, as distinguished from tidelands, which are attached to the mainland or an island and cover and uncover with the tide. Tidelands presuppose a high-water line as the upper boundary, whereas submerged lands do not. *Shoreline*—Shorelines represent the intersection of the land with the water surface. The shoreline shown on NOAA's charts represents the line of contact between the land and a selected water elevation.
These existing themes are proposed to be removed because, according to the FGDC, the data associated with them are primarily statistical.	*Cultural and Demographic Statistics*—These geospatially referenced data describe the characteristics of people, the nature of the structures in which they live and work; the economic and other activities they pursue; the facilities they use to support their health, recreational, and other needs; the environmental consequences of their presence; and the boundaries, names, and numeric codes of geographic entities used to report the information collected. *Law Enforcement Statistics*—Law enforcement statistics describe the occurrence of events (including incidences, offenses, and arrests) geospatially located, related to ordinance and statutory violations and the individuals involved in those occurrences. This theme also includes data related to deployment of law enforcement resources and performance measures. *Public Health*—Public health themes relate to the protection, improvement, and promotion of the health and safety of all people. For example, public health databases include spatial data on mortality and natality events, infectious and notifiable diseases, incident cancer cases, behavioral risk factor

FGDC proposed theme and description	Corresponding A-16 existing theme and description
	and tuberculosis surveillance, hazardous substance releases and health effects, hospital statistics, and other similar data.

Source: OMB and FGDC documentation.

[a]Designates a framework theme identified in OMB Circular No. A-16 as critical for many geospatial applications.

End Notes

[1] Department of the Interior, Geospatial Line of Business Capital Asset Summary, Aug. 14, 2012.

[2] Duplication occurs when two or more agencies or programs are engaged in the same activities or provide the same services to the same beneficiaries.

[3] GAO, 2012 Annual Report: Opportunities to Reduce Duplication, Overlap and Fragmentation, Achieve Savings, and Enhance Revenue, GAO-12-342SP (Washington, D.C.: Feb. 28, 2012); and Opportunities to Reduce Potential Duplication in Government Programs, Save Tax Dollars, and Enhance Revenue, GAO-11-318SP (Washington, D.C.: Mar. 1, 2011).

[4] Data themes are comprised of one or more sets of geospatial data that have national significance, as established by federal guidance, such as hydrography (i.e., surface water features, such as lakes, ponds, streams, and rivers).

[5] The data collected as part of the geodetic control theme are used to establish the precise location of geospatial data.

[6] This theme includes surface water features, such as lakes, ponds, streams, rivers, canals, oceans, and coastlines.

[7] This theme includes both physical and nonphysical components representing all modes of travel that allow the movement of goods and people between locations.

[8] The Bureau of the Budget became OMB in 1970.

[9] Executive Order No. 12906, Coordinating Geographic Data Acquisition and Access: The National Spatial Data Infrastructure, 59 Fed. Reg. 17,671 (Apr. 11, 1994).

[10] OMB, Circular No. A-16, Coordination of Geographic Information and Related Spatial Data Activities, Aug. 19, 2002.

[11] According to FGDC officials, there are seven framework themes, with two of the themes having two parts.

[12] http://www.geoplatform.gov.

[13] http://geo.data.gov/geoportal/catalog/main/home.page.

[14] The repository can also be searched using data.gov; however, the site automatically redirects the user to geo.data.gov. Data.gov was initially launched in May 2009 to encourage open government by increasing access to official sources of federal data and facilitating the use and development of tools to manipulate the data by all levels of government as well as academia and the private sector.

[15] OMB, M-11-03, Issuance of OMB Circular A-16 Supplemental Guidance, Nov. 10, 2010.

[16] The chair and vice-chair may select designees to serve on their behalf. The Secretary of the Interior has delegated the committee chair responsibility to the Assistant Secretary for

Water and Science. The Deputy Director for Management of OMB has delegated the committee vice-chair responsibility to the Federal Deputy Chief Information Officer.

[17] The 32 agency members of the Steering Committee are: Interior, OMB, U. S. Department of Agriculture, Commerce, Department of Defense, U.S. Army Corp of Engineers (non-voting member), Department of Education, Department of Energy, Department of Health and Human Services, Department of Homeland Security, Department of Housing and Urban Development, Department of Justice, Department of Labor, Department of State, Transportation, Department of the Treasury, Department of Veterans Affairs, Environmental Protection Agency, Federal Communications Commission (non-voting member), General Services Administration, Library of Congress, National Aeronautics and Space Administration, National Archives and Records Administration, National Capital Planning Commission (non-voting member), National Science Foundation, U.S. Nuclear Regulatory Commission, Office of Personnel Management, Small Business Administration, Smithsonian Institution, Social Security Administration, Tennessee Valley Authority, and U.S. Agency for International Development.

[18] OMB, M-06-07, Designation of a Senior Agency Official for Geospatial Information, Mar. 3, 2006, calls for select agencies to appoint to the Steering Committee policy-level officials—a chief information officer or a senior official at the assistant secretary level.

[19] The Secretary created the committee as a federal advisory committee under the Federal Advisory Committee Act.

[20] 40 U.S.C § 11101 et seq.

[21] Pub. L. No. 107-347 (Dec. 17, 2002).

[22] GAO, Geospatial Information: Better Coordination Needed to Identify and Reduce Duplicative Investments, GAO-04-703 (Washington, D.C.: June 23, 2004).

[23] GAO, Information Technology: OMB Needs to Improve Its Guidance on IT Investments, GAO-11-826 (Washington, D.C.: Sept. 29, 2011).

[24] According to OMB, a key geospatial dataset is (1) used by multiple agencies or with agency partners such as state, tribal, and local governments; (2) applied to achieve presidential priorities as expressed by OMB; (3) needed to meet shared mission goals of multiple federal agencies; or (4) expressly required by statutory mandate.

[25] OMB, Circular No. A-16, Coordination of Geographic Information and Related Spatial Data Activities, Aug. 19, 2002; and Executive Order No. 12906, Coordinating Geographic Data Acquisition and Access: The National Spatial Data Infrastructure, 59 Fed. Reg. 17,671 (Apr. 11, 1994).

[26] FGDC, FGDC-STD-001-1998: Content Standard for Digital Geospatial Metadata, 1998.

[27] The International Organization for Standardization is an international standards body that develops voluntary standards through global consensus.

[28] OMB, M-11-03, Issuance of OMB Circular A-16 Supplemental Guidance, Nov. 10, 2010.

[29] According to Interior officials, the actual number of datasets varies over time based on, for example, new technologies, new data collection, data integration, and data disposition.

[30] OMB, Circular No. A-16, Coordination of Geographic Information and Related Spatial Data Activities, Aug. 19, 2002; and Executive Order No. 12906, Coordinating Geographic Data Acquisition and Access: The National Spatial Data Infrastructure, 59 Fed. Reg. 17,671 (Apr. 11, 1994).

[31] OMB, Circular No. A-11, Preparation, Submission, and Execution of the Budget, section 25, Aug. 3, 2012; and Executive Order No. 12906, Coordinating Geographic Data Acquisition and Access: The National Spatial Data Infrastructure, 59 Fed. Reg. 17,671 (Apr. 11, 1994).

[32] As previously noted, metadata are information about datasets, such as content, source, accuracy, method of collection, and point of contact.

[33] According to an FGDC Secretariat official, the Geospatial One-Stop had previously included the ability for agencies to identify their planned acquisitions. This official told us that during the time the Geospatial One-Stop was available, there were approximately 4,000 instances of organizations notifying the geospatial community, through the One-Stop, of planned data acquisitions.

[34] OMB, Circular No. A-16, Coordination of Geographic Information and Related Spatial Data Activities, Aug.19, 2002.

[35] 5 U.S.C. § 306(a)(1), (2), (4), (6), and (7).

[36] OMB, Circular No. A-11, Preparation, Submission, and Execution of the Budget, section 210, Aug. 3, 2012.

[37] OMB, M-06-07, Designation of a Senior Agency Official for Geospatial Information, Mar. 3, 2006; OMB, Circular No. A-16, Coordination of Geographic Information and Related Spatial Data Activities, Aug.19, 2002; and Executive Order No. 12906, Coordinating Geographic Data Acquisition and Access: The National Spatial Data Infrastructure, 59 Fed. Reg. 17,671 (Apr. 11, 1994).

[38] Department of the Interior, Geospatial Modernization Blueprint Recommendations and Architectures, Version 1.0, December 2007.

[39] National Oceanic and Atmospheric Administration, Environmental Data Management Committee, Data Documentation Procedural Directive, Oct. 28, 2011.

[40] Census Bureau, National State Geographic Partnerships Branch, Geospatial Product Metadata Standard, Version 5.0, Jan. 28, 2010.

[41] The number of federal records in the clearinghouse as of February 2012 (441,343) is higher than the total number of federal records in the clearinghouse as of August 2012 (372,986) because, according to the General Services Administration, who is responsible for managing the clearinghouse database, it has taken steps to remove duplicative records or records no longer owned by federal agencies.

[42] According to the FGDC, Content Standard for Digital Geospatial Metadata Workbook, May 1, 2000, there are 29 mandatory fields that are to be completed for all metadata records.

[43] Elevation data provide three-dimensional models of the Earth's surface. Orthoimagery data provide images of the Earth's surface collected by aerial photography or satellites.

[44] OMB, Circular No. A-16, Coordination of Geographic Information and Related Spatial Data Activities, Aug.19, 2002.

[45] Department of Commerce, National Oceanic and Atmospheric Administration, The National Geodetic Survey Ten-Year Plan: Mission, Vision, and Strategy 2008-2018, 2008.

[46] The geometric datum, also known as the horizontal datum, is the horizontal frame of reference used by all civilian agencies. It is essentially the x and y axis on a horizontal plane, similar to longitude and latitude.

[47] The geopotential datum, also known as the vertical datum, is the vertical frame of reference used by all civilian agencies. It is essentially the z axis in a three-dimensional model, similar to altitude.

[48] Road centerlines are vector line data that represent the geographic center of road rightsof-way on transportation networks.

[49] The FGDC Secretariat provided a list of key datasets associated with the three themes in our sample as of August 2012.

[50] The key dataset was the Continuously Operating Reference Stations.

[51] The key datasets were National Hydrography Dataset, National Water Information System-Stream Gage Locations, National Water Information System-Water Quality Monitoring Stations Locations, and National Wetlands Inventory.

[52] The key dataset was the National Hydrography Dataset.

[53] As of August 2012, NOAA officials stated that they had requested a modification to its exhibit 53 to identify the Continuously Operating Reference Stations dataset as geospatial services.

[54] "Leaf-on" imagery refers to the time of year in which imagery data are being acquired during which there is foliage on certain tree and shrub species.

[55] "Leaf-off" imagery refers to the time of year in which imagery data are being acquired during which there is no foliage or a reduced amount of foliage on certain tree and shrub species.

[56] Department of Transportation, Transportation for the Nation Strategic Plan, May 2011.

[57] According to Interior officials, the Census Bureau and USGS have formed a working group to determine if the accuracy of the TIGER data can be improved sufficiently to allow them to be used on the National Map website, as opposed to the licensed data.

[58] The report is tentatively scheduled for approval by the National Geospatial Advisory Committee by December 2012.

[59] Founded by congressional charter, the National Academy of the Sciences is a private, nonprofit organization that serves as advisors to the nation on issues of science and technology that frequently affect policy decisions.

[60] National Academy of Sciences, National Land Parcel Data: A Vision for the Future, 2007.

[61] Congressional Research Service, Issues and Challenges for Federal Geospatial Information, Apr. 27, 2012; and Issues Regarding a National Land Parcel Database, May 13, 2011.

[62] The National States Geographic Information Council has two missions: (1) to promote the coordination of statewide geospatial activities in all states and (2) to advocate for the states in national geospatial policy initiatives to help enable the NSDI. Members of the council include senior state GIS managers and coordinators as well as others from all levels of government, academia, and the private sector. See http://www.nsgic.org/.

End Notes for Appendix I

[1] The FGDC considers there to be seven framework themes, with two of the themes having two parts. Specifically, FGDC considers cadastral and cadastral (offshore) to be one theme and elevation bathymetric and elevation terrestrial to be one theme.

[2] Executive Order No. 12906, Coordinating Geographic Data Acquisition and Access: The National Spatial Data Infrastructure, 59 Fed. Reg. 17,671 (Apr. 11, 1994).

[3] OMB, Circular No. A-16, Coordination of Geographic Information and Related Spatial Data Activities, Aug. 19, 2002.

[4] OMB, M-06-07, Designation of a Senior Agency Official for Geospatial Information, Mar. 3, 2006.

[5] OMB, M-11-03, Issuance of OMB Circular A-16 Supplemental Guidance, Nov. 10, 2010.

[6] FGDC, FGDC-STD-001-1998: Content Standard for Digital Geospatial Metadata, 1998; and Content Standard for Digital Geospatial Metadata Workbook, May 1, 2000.

[7] OMB, Circular No. A-11, Preparation, Submission, and Execution of the Budget, August 2011; and Circular No. A-11, Preparation, Submission, and Execution of the Budget, August 2012.

[8] See, for example: Department of Transportation, Transportation for the Nation Strategic Plan, May 2011; National Geospatial Advisory Committee, National Addresses Databases Background Paper - DRAFT, prepared by Gene Trobia, Apr. 3, 2012; National Academy of Sciences, National Land Parcel Data: A Vision for the Future, 2007; Congressional Research Service, Issues and Challenges for Federal Geospatial Information, Apr. 27, 2012; and Congressional Research Service, Issues Regarding a National Land Parcel Database, May 13, 2011.

[9] The committee is a federal advisory committee sponsored by Interior under the Federal Advisory Committee Act. It reports to the Chair of the FGDC.

[10] Established by congressional charter, the National Academy of Sciences is a private, nonprofit organization that serves as advisors to the nation on issues of science and technology that frequently affect policy decisions.

[11] The National States Geographic Information Council is made up of senior state geographic information system managers and coordinators. Other members include representatives from federal agencies, local government, the private sector, academia, and other professional organizations.

In: Geospatial Data
Editors: Meredith Haynes

ISBN: 978-1-63483-524-4

Chapter 3

NATIONAL SPATIAL DATA INFRASTRUCTURE STRATEGIC PLAN 2014–2016*

Federal Geographic Data Committee

EXECUTIVE SUMMARY

The tremendous growth and evolution of geospatial technologies, industries, and applications in recent years has highlighted the need to update and modernize the strategic direction of Federal geospatial programs to align with and leverage these advancements. The Federal Geographic Data Committee's (FGDC) 2014–2016 Strategic Plan for the National Spatial Data Infrastructure (NSDI) sets priorities and describes the actions the FGDC community will take, in collaboration with partners, to develop and maintain the Nation's critical geospatial infrastructure.

This Strategic Plan was developed during 2013 through extensive consultation with Federal agencies, the National Geospatial Advisory Committee (NGAC), and with partners and stakeholders across the geospatial community. This external input was instrumental in shaping the plan and in reflecting the perspectives of the broader geospatial community.

This plan describes a shared national vision of the NSDI and includes a set of goals and objectives for the role of Federal agencies in achieving this vision:

---

* This is an edited, reformatted and augmented version of a report issued December 2013.

The NSDI leverages investments in people, technology, data, and procedures to create and provide the geospatial knowledge required to understand, protect, and promote our national and global interests.

The goals and objectives in the plan define areas of critical importance to the continued development of the NSDI and provide a framework for the Federal geospatial community's responsibilities over the next 3 years.

- Goal 1 – Develop Capabilities for National Shared Services
  - Objective 1.1. Develop geospatial interoperability reference architecture.
  - Objective 1.2. Establish the Geospatial Platform as the Federal geospatial data, services, and applications Web-based service environment.
  - Objective 1.3. Expand the use of cloud computing.
  - Objective 1.4. Promote the use of geospatial multiagency acquisition vehicles for interagency and intergovernmental purchases.

- Goal 2 – Ensure Accountability and Effective Development and Management of Federal Geospatial Resources
  - Objective 2.1. Advance the portfolio management process for National Geospatial Data Assets (NGDA).
  - Objective 2.2. Identify potentially duplicative investments and opportunities for collaborative investments.

- Goal 3 – Convene Leadership of the National Geospatial Community
  - Objective 3.1. Lead and participate in the development and coordination of national and international standards applicable to the geospatial community.
  - Objective 3.2. Convene the leadership of the geospatial and nongeospatial communities to develop shared approaches to critical national issues.
  - Objective 3.3. Raise awareness of the NSDI and its impact on critical national issues.

The Strategic Plan also describes the steps the FGDC community will take to implement the goals, objectives, and actions, including development of project plans and performance measures in collaboration with partners and stakeholders. For more information, visit www.fgdc.gov/nsdi-plan.

THE NATIONAL SPATIAL DATA INFRASTRUCTURE

Executive Order 12906, "Coordinating Geographic Data Acquisition and Access," describes the National Spatial Data Infrastructure (NSDI) as "the technology, policies, standards, and human resources necessary to acquire, process, store, distribute, and improve utilization of geospatial data."

A Shared Vision for the NSDI

The FGDC has worked with partners and stakeholders, including the members and organizations represented on the National Geospatial Advisory Committee, to collaboratively define a shared national vision that describes the value that the NSDI will bring to its stakeholders and partners.

NSDI Vision Statement

The NSDI leverages investments in people, technology, data, and procedures to create and provide the geospatial knowledge required to understand, protect, and promote our national and global interests.

The National Spatial Data Infrastructure extends far beyond data. The NSDI encompasses the policies, organizational responsibilities, data, information, technologies, standards, services, and financial and human resources necessary to achieve this vision. The NSDI has become a critical vehicle for facilitating seamless data development, information sharing, and collaborative decisionmaking across multiple sectors of the economy. The following describes the vision of the NSDI for the future.

The Desired Future State of the NSDI

- Provide government, businesses, and citizens with a way to visualize and explore data to derive information and knowledge.
- Create a network of resources and services for the seamless integration of location-based information into broader information assets to serve the

needs of government, the business community, and citizens.
- Serve as an enabling resource for discovery, access, integration, and application of location information for a growing body of users.
- Leverage shared and open standards-based services and focus on applied information for improved decisionmaking.
- Promote place-based business intelligence and smart, shared applications.
- Include a core set of information layers that interface with other nonspatial data being generated.
- Use real-time data feeds and sensor webs for improved monitoring, control, situational awareness, and decisionmaking.
- Facilitate access to and use of multi-temporal information linked to place.
- Integrate and use advanced technologies and their associated standards and best practices.
- Facilitate use of community-driven open standards with multiple implementations.

GUIDING PRINCIPLES

Federal agencies that produce, collect, maintain, or use spatial data either directly or indirectly will (1) recognize and manage their data as capital assets, (2) facilitate non-Federal participation in the development of the NSDI, and (3) work together through the FGDC to provide for effective and economical use and management of geospatial data in the digital environment for the benefit of the Nation.

Guiding Principles for the Federal Geospatial Community

- Ensure that spatial data from multiple sources (Federal, State, Tribal, regional, and local governments; academia; and the private sector) are available and easily integrated to enhance understanding of our physical, natural, and cultural world.
- Facilitate the development of authoritative National Geospatial Data Assets that are complete, accurate, current, standards-compliant, and at the scale needed for shared uses by Federal, State, Tribal, regional, and local governments, academia, the private sector, and the public.

- Protect the privacy and security of citizens' personal data and ensure the accuracy of statistical information about people, both in raw form and in derived information products.
- Enable access for all citizens to spatial data, information, and derivative and interpretive products, in accordance with Office of Management and Budget Circulars A–130 and A–16 and the Open Data Policy.
- Protect proprietary interests related to licensed information and data.
- Enable interoperability of information systems through the use of open and machine-readable formats to enable access to resources from multiple agencies and partners.
- Ensure that investment and policy decisions consider the expected return on investment and effective use of resources.

STRATEGIC GOALS

The NSDI Strategic Plan includes a set of three strategic goals that were developed through extensive consultation with Federal agencies and partners and stakeholders across the geospatial community. The strategic goals include objectives and actions that describe how the goals will be accomplished. The following section describes the strategic goals, objectives and actions in the Strategic Plan; and also describes the outcomes that will result from their implementation.

GOAL 1 – DEVELOP CAPABILITIES FOR NATIONAL SHARED SERVICES

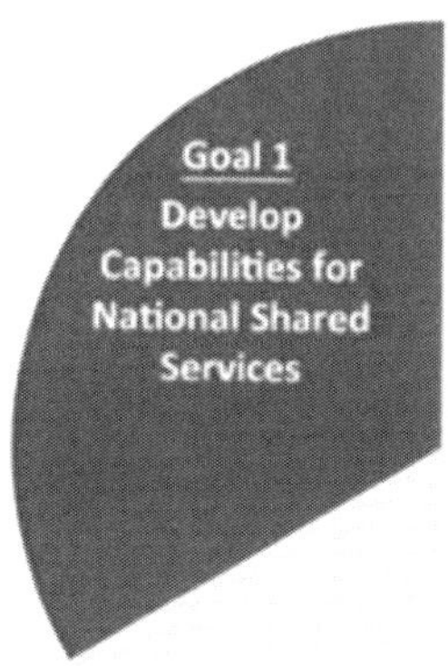

Developing shared services is a vital strategy for delivering geospatial solutions faster, for less money, and with fewer resources. These enterprise approaches take advantage of evolving technologies and methodologies to reduce potential duplication, create innovative tools and technology, and leverage agency budgets and expertise to accomplish agency mission and support functions more efficiently, while also improving quality.

The Geospatial Platform initiative is a critical component for the continued development of the NSDI. The Platform is a Web-based first generation service environment that provides access to a suite of well-managed, highly available, and trusted geospatial data, services, applications, and tools for use by Federal agencies and their State, Tribal, regional, and local partners. In addition, the FGDC community and its partners will utilize common cloud computing and enterprise acquisition approaches as mechanisms to leverage technology, close productivity gaps, and combine buying power across agencies for similar needs.

This strategic goal describes how the Federal geospatial community will work with partners to develop shared service approaches and leverage the Administration's Federal Information Technology Shared Services Strategy.

Objective 1.1. Develop Geospatial Interoperability Reference Architecture

Anticipated Outcomes. More current, complete data available to all through improving capabilities for NSDI partners to develop, maintain, share, and use geospatial information. Cost savings through better leveraging of shared technology infrastructure across Federal agencies and among our partners.

- Action 1.1.1. Establish reference architecture to assure interoperability utilizing published open service standards to share data among unclassified, controlled unclassified information (CUI), and classified domains and missions.
- Action 1.1.2. Define the role of the Geospatial Platform as a technology and collaboration environment for unclassified geospatial information sharing.
- Action 1.1.3. Identify Federal Chief Information Officer (CIO) Council requirements, shared services, and other infrastructure that can be reused and leveraged by the NSDI, including access control, search, and discovery.

Objective 1.2. Establish the Geospatial Platform as the Federal Geospatial Data, Services, and Applications Web-Based Service Environment

Anticipated Outcomes. Efficient, effective use of shared technology infrastructure to encourage and support better information sharing. Cost savings through shared geospatial investment planning.

- *Action 1.2.1.* Establish Service Level or Interagency Agreements between and among government agencies and the Geospatial Platform Managing Partner.
- *Action 1.2.2.* Establish the Geospatial Platform Oversight Body and develop its operating procedures, scope, and roles for Federal and non-Federal members.
- *Action 1.2.3.* Implement the primary contracting mechanism to continue Geospatial Platform development and operations and maintenance for FY 2014 and beyond.
- *Action 1.2.4.* Implement communities of interest on the Geospatial Platform for collaboration, including a shared investment planning "Marketplace" and data theme communities, as outlined in OMB Circular A–16.
- *Action 1.2.5.* Develop guidance for Federal agencies and their partners that describes how to use the Geospatial Platform to store and publish data and metadata and how to use services in the Platform's common hosting infrastructure.
- *Action 1.2.6.* Solicit advice and recommendations from the NGAC and community partners on ways to expand and enhance the Geospatial Platform for broader use and value by the non-Federal community.
- *Action 1.2.7.* Define the concept and develop an implementation plan for a "Data as a Service" offering within the Geospatial Platform Marketplace.

Objective 1.3. Expand the Use of Cloud Computing

Anticipated Outcomes. Cost savings through expanded use of commodity cloud-hosting capabilities for the benefit of all NSDI partners.

- *Action 1.3.1.* Define the mechanism and workflow for Geospatial Platform leveraging of a Department of the Interior (DOI) Foundation Cloud Services offering that will benefit the Platform user community.
- *Action 1.3.2.* Provide guidance, best practices, and case studies for agencies considering migration of agency-stored content and services to commodity cloud providers.

- *Action 1.3.3.* Develop an options paper for expanding the currently defined common hosting environment to look at other capabilities available in the commercial market.

Objective 1.4. Promote the Use of Geospatial Multiagency Acquisition Vehicles for Interagency and Intergovernmental Purchases

Anticipated Outcomes. Cost savings through better leveraging of governmental purchasing power when buying geospatially related goods and services for the benefit of multiple partners.

- *Action 1.4.1.* Inventory available and planned geospatial acquisition vehicles open to Federal agencies and non-Federal partners.
- *Action 1.4.2.* Create a mechanism for sharing information on the availability and use of consolidated acquisition vehicles.

GOAL 2 – ENSURE ACCOUNTABILITY AND EFFECTIVE DEVELOPMENT AND MANAGEMENT OF FEDERAL GEOSPATIAL RESOURCES

Effective management of geospatial investments will enable Federal agencies and their partners to meet mission responsibilities, control costs, streamline services, make decisions on eliminating duplication, save taxpayer dollars, and drive efficiency across the Federal Government. The OMB Circular A–16 Supplemental Guidance provides guidelines for implementing a portfolio management approach to Federal geospatial investments and managing National Geospatial Data Assets (NGDA) governmentwide. This approach will allow identification of data themes and datasets that are most critical for meeting the needs of government and stakeholders. The FGDC agencies will also apply the framework of the Administration's Open Data Policy to manage information as an asset throughout its life cycle to promote interoperability and openness and safeguard systems and information.

This strategic goal describes the actions the Federal geospatial community will take to implement portfolio management to more effectively plan geospatial data collection efforts, assess the status of NGDAs, and minimize duplicative investments.

Objective 2.1. Advance the Portfolio Management Process for National Geospatial Data Assets (NGDA)

Anticipated Outcomes. Increased access to high-priority, documented, and standards-based national geospatial datasets and services. Enhanced level of transparency into Federal activities and resources required to manage and provide access to geospatial resources.

- *Action 2.1.1.* Identify foundational NGDAs and high-priority data themes and datasets to be included in NSDI portfolio management.
- *Action 2.1.2.* Define Federal roles and responsibilities in national data management, including metadata and data delivery, taking into account OMB Circular A–16 Supplemental Guidance portfolio management requirements, the Open Data Policy, Geospatial Platform, Data.gov, and other relevant requirements.
- *Action 2.1.3.* Finalize and implement the Circular A–16 Portfolio Implementation Plan to include reporting investments and defining investment requirements.
- *Action 2.1.4.* Develop a process for monitoring and reporting on the progress of Circular A–16 Data Theme and Geospatial Platform

Community management responsibilities, including the use and proliferation of content and technology standards.

Objective 2.2. Identify Potentially Duplicative Investments and Opportunities for Collaborative Investments

Anticipated Outcomes. Additional information is publicly available on geospatial acquisition plans resulting in cost efficiencies and decreased duplication of effort. Enhanced return on geospatial investments through partnerships across Federal and non-Federal sectors.

- *Action 2.2.1.* Provide guidance and instructions to Federal agencies on use of the Geospatial Platform Marketplace.
- *Action 2.2.2.* Develop process and technology implementation to track use of the Geospatial Platform Marketplace and cost savings/avoidance through its application.
- *Action 2.2.3.* Solicit feedback from the Geospatial Platform user community on future requirements for Marketplace functionality.

GOAL 3 – CONVENE LEADERSHIP OF THE NATIONAL GEOSPATIAL COMMUNITY

The Federal Government is in a unique position to provide a leadership and facilitation role in the national geospatial community. This role includes providing institutional leadership for the development and coordination of national and international geospatial standards and policies, integrating geospatial technology into information technology (IT) management processes, contributing to the development of a legal and policy framework that supports the objectives of the NSDI, and fostering collaboration across sectors. The Federal geospatial community can play an important role in communicating the value of geospatial information and tools to enable informed analysis and decisionmaking. Stakeholders have identified this strategic goal as a critical responsibility that can be accomplished most effectively by the Federal Government.

Objective 3.1. Lead and Participate in the Development and Coordination of National and International Standards Applicable to the Geospatial Community

Anticipated Outcomes. Greater adoption and utilization of standards resulting in enhanced interoperability of geospatial data, services, and systems. Decreased barriers to exchanging geospatial content effectively and efficiently by reducing technical impediments to sharing geospatial data and services.

- *Action 3.1.1.* Consult and collaborate with both existing and emerging geospatial communities to advance common standards and approaches.
- *Action 3.1.2.* Develop strategic partnerships with existing standards development organizations.
- *Action 3.1.3.* Provide technical and educational support to agencies implementing NSDI standards and engage the standards community through workshops or online information-sharing tools such as Web meetings or social media.

Objective 3.2. Convene the Leadership of the Geospatial and Nongeospatial Communities to Develop Public/Private Partnerships and Shared Approaches for Addressing Critical National Issues

Anticipated Outcomes. More effective collaboration in the development and oversight of critical geospatial activities required for the effective management of the NSDI.

- *Action 3.2.1.* Engage with the NGAC, key geospatial organizations, and other stakeholders to inform policy decisions and collaboratively identify and address issues of common concern.
- *Action 3.2.2.* Collaborate with the NGAC and other partners to review and develop common approaches to the issue of geolocation privacy.

Objective 3.3. Raise Awareness of the NSDI and Its Impact on Critical National aqnd International Issues

Anticipated Outcomes. Enhanced level of understanding, engagement, and investment by all affected communities in the development and maintenance of the national spatial data infrastructure.

- *Action 3.3.1.* Develop and implement a communication strategy and outreach plan to promote the benefits of NSDI data and the goals of the NSDI Strategic Plan.
- *Action 3.3.2.* Promote the vision of NSDI globally through active participation and support for activities such as the Global Spatial Data Infrastructure (GSDI), the United Nations Initiative on Global Geospatial Information Management (UN-GGIM), and the U.S. Group on Earth Observations (USGEO).

STRATEGIC PLAN OVERVIEW

Background

The Changing Geospatial Landscape

The technology landscape has seen tremendous change since the NSDI was initially conceptualized in the early 1990s. In the years leading up to the 1990s, advances in mapping and geospatial technology were largely driven by the public sector—with Federal agencies playing a major role in the development and coordination of data, products, and services. The origins for many of the current geospatial innovations and operational successes have their roots in past government investment in research and implementation of computing and communication technology, geospatial data standards, NSDI framework data development, the Internet, remote sensors, and the Global Positioning System (GPS).

In the years since the inception of the NSDI, remarkable advances in computing power, the emergence of open standards and open systems, and the Internet have reshaped the technology landscape. In recent years, the availability and usage of geospatial information and products have dramatically expanded as software has matured and high-quality data became available through newly implemented State and Federal clearinghouses and private sector offerings.

The Growing Role and Value of Geospatial Technology and Information

The geospatial technology and services industry is a growing and important factor in the United States and world economies, driving significant benefits and providing high-wage jobs. A 2012 study by the Boston Consulting Group (BCG) estimated that the U.S. geospatial industry generated approximately $73 billion in revenues in 2011 and comprises at least 500,000 high-wage jobs. In addition, BCG found that geospatial services deliver efficiency gains in the rest of the economy that are valued at many times the size of the sector itself—with geospatial services driving $1.6 trillion in revenue and $1.4 trillion in cost savings. These benefits, representing 15 to 20 times the size of the geospatial services sector itself, create an important competitive advantage for the U.S. economy. In addition, the U.S. Department of Labor recently identified the geospatial technology sector as one of the three technology areas that would create the greatest number of new jobs over the next decade.

Worldwide, a 2012 study by Oxera commissioned by Google estimated that the global geospatial services sector generates $150 to $270 billion annually. By comparison, this is greater than the $25 billion generated by the video games industry, roughly equivalent to the $140 billion in revenue from the global security services industry, and about one-third of the global airline industry's annual revenues of $594 billion.

The Role of the FGDC

The Office of Management and Budget (OMB) has specific oversight responsibilities for Federal information technology systems and acquisition activities, including geographic information systems, to help ensure their efficient and effective use. Since 1953, OMB has released several iterations of Circular A–16 that, among other requirements, encourages avoidance of duplicative mapping and surveying efforts. These efforts were reinforced in 1994 by Executive Order 12906, "Coordinating Geographic Data Acquisition and Access." Two subsequent laws that reinforce these responsibilities are the Clinger-Cohen Act of 1996 (in Public Law 104–106) and the E–Government Act of 2002 (Public Law 107–347).

The FGDC, which was formally established by OMB in 1990, is the interagency committee that promotes the coordinated use, sharing, and dissemination of geospatial data in the United States. The Secretary of the Interior is the Chair of the FGDC and the Deputy Director for Management of OMB is the committee's Vice-Chair. The Department of the Interior provides support for the FGDC through the Office of the Secretariat, which is housed within the U.S. Geological Survey.

For the past two decades the FGDC has worked to develop policies and partnerships to advance the development of the NSDI. To achieve this, the FGDC has contributed to the evolution of Federal and national geospatial initiatives. Several of these initiatives have been sponsored as Administration priorities and are highlighted in the following figure:

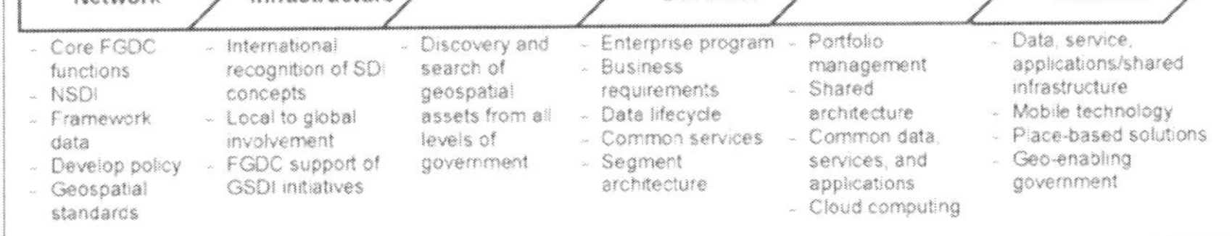

Most recently, the FGDC led the development of the OMB Circular A–16 Supplemental Guidance, initiated the implementation of a geospatial portfolio management approach, and led the development of the Geospatial Platform initiative. The OMB Circular A–16 Supplemental Guidance addresses portfolio management for the coordination of Federal geospatial data assets and investments to more efficiently support national priorities and government missions. Portfolio management will apply consistent management approaches to help increase the quality of data through best practices and documentation in a manner that reduces duplication and cost and provides greater accessibility. The Geospatial Platform will provide a suite of shared services and capabilities to Federal agencies and partners. These developments build upon the previous processes and accomplishments of the FGDC and will provide a means to more efficiently manage and deliver Federal geospatial products and services in the coming years. The Federal activities in the Strategic Plan will be carried out by the FGDC community, which includes FGDC member agencies, the Steering Committee, Coordination Group, subcommittees, working groups, and partners.

Strategic Plan Drivers

This Strategic Plan is timely for several reasons. First, while the FGDC has engaged in several recent strategic initiatives, including the Geospatial Line of Business and the Geospatial Platform, the current NSDI Strategic Plan has not been revised since 2005. Secondly, geospatial technologies, industries, and applications have seen tremendous growth and change over the past 8 years: therefore; strategies need to be updated and modernized to align with and leverage these changes.

In addition, the U.S. Government Accountability Office (GAO) published "Geospatial Information - OMB and Agencies Need to Make Coordination a Priority to Reduce Duplication" (GAO-13-94) in November 2012. In the report, GAO recommended that the Secretary of the Interior, as the FGDC chair, should establish a time frame for creating and updating a strategic plan to improve coordination and reduce duplication, and create and implement the plan within the established time frame. This Strategic Plan fulfills the requirement under OMB Circular A–16 to "Prepare and maintain a strategic plan for the development and implementation of the NSDI.

Roles and Responsibilities of the Federal Sector

The FGDC, in collaboration with partners and with input from the National Geospatial Advisory Committee (NGAC), has examined the current and future roles of the Federal Government as they relate to the continued development of the NSDI. The following is a summary of the key geospatial roles for the Federal Government and relationships with other sectors of the geospatial community.

Role	Responsibilities
Leadership	The Federal Government will play a critical role in providing institutional leadership for the development and coordination of national and international geospatial policies, the alignment of geospatial technology with information management and IT processes, and the development of a legal and policy framework that supports the objectives of the NSDI. Federal agencies are in a unique position to convene multiple stakeholders, through mechanisms such as the NGAC, to develop joint approaches to issues affecting multiple sectors. In addition, the Federal geospatial community can provide leadership to establish joint acquisitions of geospatial data and services (for example, SmartBUY) with buy-up options for partners.
Partner	Federal agencies will continue to work together and with State, Tribal, regional, local, academic, and private sector partners to jointly fund and share geospatial data, services, and tools. Emphasis will be placed on improved efficiency, improved effectiveness, and open access as directed by the Administration's Open Data Policy.
Facilitator	The FGDC will facilitate collaboration between and across all levels of government, academia, nonprofits, the private sector, and individuals for research and to create data and information services. Pursuant to this role, Federal agencies can convene key experts, decisionmakers, and stakeholders to address common concerns and solve problems. The Federal role will be to work with other levels of government and stakeholders to share data, approaches, and services; to develop best practices to meet common needs; and to clearly delineate responsibilities for agency theme leads under OMB Circular A–16.
Provider	Federal agencies will focus on providing geospatial services and tools for the use of critical data assets. This focus encompasses the agency mapping activities, framework data, the A–16 categories, metadata and catalogs, and interagency communications. The Federal geospatial community should focus on the implementation of the Geospatial Platform as a primary mechanism for collaborative development and distribution of data content via standards-based services.
Curator	Federal agencies will provide and maintain geospatial content, where appropriate, and focus on delivering services that provide data in standardized, accessible, open formats. The Federal community should take the lead in

Rôle	Responsibilities
	building and maintaining an online catalog for transparent discovery and access to Federal and other data through the Geospatial Platform and other resources.
Enabler	The Federal Government will play an enabler role to make geospatial data and services more accessible to an expanded audience through tools such as the Geospatial Platform. Federal agencies will work with partners to develop more timely, accurate, and useful data and services. In this way the Federal Government can make State and local information play an enhanced role in national government and business operations.
Standards and Research	The Federal Government will promote the continued development of consensus-based standards for interoperability as new technologies and capabilities evolve in order to achieve and maintain an interoperable NSDI network. In addition, the Federal Government will effectively coordinate among programs that conduct and (or) sponsor geospatial research to ensure that highest priority needs are addressed; the Federal Government will work with partners to better share research results and translate new findings and capabilities into operational activities across appropriate sectors.

EXTERNAL FACTORS

The National Geospatial Advisory Committee (NGAC) has identified the following key factors, including technology, economic, political, and organizational trends, that should be considered for geospatial strategy at the national level. The goals and objectives outlined in this plan will address the key issues identified by NGAC.

Evolutions in Information Management

Innovations in data collection, management, processing, dissemination, and application are becoming available largely through private sector advancement of sensors, crowd-sourcing, big data storage/analytics and information distribution technologies, and through academic and government funded research. In light of these changes, the role of government is transitioning from building framework spatial infrastructure to ensuring that specific programmatic goals and the greater public good are served by trusted, documented information resources. This information has new requirements for transparency, currency, accuracy, and accessibility. The next generation of geospatial resources will be information sourced and served from a combination of the private sector, the public sector, individuals, nongovernment groups, and public-private partnerships.

Geospatial information is now viewed as a key component of the overall information management fabric. Geospatial infrastructure provides value via connected services to a diverse set of devices operated by end users comprising a full range of roles and skills. It balances the same demands as other information resources (accessibility, security, persistence, reliability, confidentiality, and cost efficiency), while providing the spatial capabilities that are its core strength: visualization, analytics, and the ability to bring together disparate information resources using location as a unifying principle.

Important Trends

Other trends impacting the strategy for a national geospatial infrastructure can be characterized by four themes: technology, communications, workforce, and legal/policy issues.

Technology

- The "Internet of Things" will connect billions of stationary and mobile sensors with human users.
- The value of integrated geospatial information (3D, 4D, indoor, unstructured, linked, archived) will be multiplied by high-powered processing and analytics capabilities, and by the ability to provide succinct intelligence to decisionmakers.
- Static ad hoc data will yield to big data transactional models based on cloud-based service models for infrastructure, software, and information content.
- The needs and practice of information technology management and geospatial management will align.
- Increasingly, geospatial information resources and application source code will be built collaboratively, using open, rapid-deployment strategies and open standards.

Impacts: Geospatial technologies will be called upon to organize a much larger information domain, provide trusted analysis of complex "big data" holdings, and effectively visualize and communicate knowledge so that it can be turned into operational efficiencies. Open data and open government initiatives built on the desire for efficiency and transparency will be the norm, and these initiatives must be balanced against traditional IT requirements for appropriate levels of information security and risk management. Geospatial strategies should use life-cycle approaches that factor in time to market and future agility in systems development and evolution.

Communications

- Systems for information delivery will be ubiquitous and highly mobile, will utilize Web-based services, and will be components of more advanced information workflows.

Traditional media will be challenged to adapt to the changes initiated by dynamic social media channels.

- Agencies and individuals will have far higher expectations for the quality, timeliness, currency, and accessibility of geospatial data and services, yet will have ubiquitous access to information with vastly different lineages and degrees of uncertainty.

Impacts: There is a heightened need to collect, store, and process data from sensors, crowd-sourcing, and other processes and connect the data to decisionmaking processes as actionable information. Harnessing content provided by sensors and social media, in particular, as feedback to enhance authoritative processes and products, needs research and strategy. Information regarding data, processes, and products must be more usable and reliable to ensure meaningful applications of data to create knowledge and inform decisions. Additional attention is needed with regard to the measurement and expression of the uncertainty inherent in information and related analytical products.

Workforce

- + There will be increasing demand for skill sets positioned at the intersection between the traditional IT and geospatial realms, such as application and services development, geoinformation fusion, crowd-sourcing, social networks and human geography, identity management, visual analytics, and forecasting/modeling.

- + There will be increasing acceptance of professional qualification from nontraditional sources: massive open online courses (MOOC) and other online training and information resources.
- + Responsibilities for geospatial data and shared services operations will increasingly be assigned to information technology and chief information officers.

Impacts: The current capabilities and organization of the geospatial workforce will be called upon to transform at a rapid pace. This is no small task, as it will require a combination of finding, retraining, reorganizing, and replacing existing staff. Successful geospatial organizations will build and maintain agile workforces closely aligned to or within IT divisions, while also retaining and supporting competencies in fundamental geographic and geomatics disciplines. IT security governance models will need to adapt to an evolving cloud deployment environment where infrastructure and platform security is easily shared across agencies.

Legal/Policy Issues

- + Views and approaches to geospatial privacy will continue to evolve. Geolocation privacy is a complex and multifaceted issue that will engage a variety of interests, including both geospatial and nongeospatial entities. Potential statutory or regulatory changes to locational privacy issues have the potential to significantly impact the geospatial community. It is likely that legislation will lag behind the changes in technology and behavior.
- + Capabilities of State, Tribal, regional, and local geospatial resources continue to improve and, in many areas, will provide the best sources for ongoing current and accurate information. Integrating these diverse datasets involves conforming to numerous legal and policy frameworks.
- + Government policy and private practices will increasingly demand information transparency. Data will be considered an integral and accountable part of decisionmaking, including its use in models and forecasts.

Impacts: Proactive approaches are needed to develop consensus terminology, develop policy, and educate citizens and decisionmakers with regard to geospatial data gathering, dissemination, licensing, and usage practices. Privacy, confidentiality, and security issues must be addressed for sensitive geospatial information, especially when directly linked to personally identifiable information, to protect the interests of both individuals and commerce. A priority for privacy policy efforts should be developing consensus on acceptable practices with immediate use. Issues at the margins of privacy acceptability should not delay important work to aggregate and utilize geospatial data and services from local to national levels in a timely manner.

Summary: What External Factors will Influence the Future of the NSDI?

A number of critical external factors are driving trends in technology, communications, workforce, and legal/policy issues of importance to a future NSDI. These factors signal a strong need to align national geospatial strategy with both mainstream and emerging information technology practices and with related trends. Geospatial technologies provide attractive integrative approaches for meeting current requirements to "do more with less." New approaches are demanded, supported, and must be implemented quickly for technology, collaboration, workforce development, funding, and other resources. This will include raising awareness of the unique aspects of geolocation from a privacy standpoint. Geospatial technology will increasingly be aligned with and integrated into the broader, ever-evolving technology ecosystem.

Challenges and Opportunities

The revolutionary changes in geospatial technology and applications and the growing visibility of geospatial technology across all sectors of the global economy are leading public sector agencies to realign their geospatial strategies and investments to meet the needs of newly geospatial-savvy information consumers. More and more, citizens expect government to provide geospatial data and services to them in their specific areas of interest. As the public sector faces the challenge of doing more with less, the smart, strategic use of shared, enterprise geospatial technologies provides significant opportunities for efficiencies. Key opportunities and challenges for the geospatial community include the following:

- The Federal Government plays an important role in providing leadership, policy development, and facilitation of the broad range of organizations and interests involved in the geospatial community. This leadership role has been identified by stakeholders as a critical responsibility.
- The geospatial community holds a shared value in the importance of standards, data sharing, partnerships, and collaboration to solve real issues. The diverse geospatial community relies on the FGDC's leadership for metadata guidance and the coordination of standards.
- Execution and enforcement of OMB Circular A–16 responsibilities have been identified as challenges for the Federal geospatial community. The task of building the NSDI and the accomplishment of A–16 responsibilities has not always been aligned with agency missions. With Federal agencies and partner organizations facing difficult budget situations, ensuring agency accountability for A–16 responsibilities will continue to be a significant challenge.
- It is difficult to quantify the entire magnitude of geospatial investment across the Federal geospatial community because these investments are typically dispersed across multiple programs and services. Moreover, geospatial data management is not fully integrated into Federal IT management processes, and many programs that have geospatial components are not visible through existing IT management and budgeting processes.
- With the expanded use of geospatial data and services, new challenges are arising for the Federal geospatial community. It has been difficult to express the value and benefit of Federal investments in providing geospatial data and services when the public can readily access a great deal of geospatial information managed by both the public and private sectors through Web-based applications and hand-held devices. In addition, technology adoption and policy development by Federal agencies often cannot keep up with the speed of the innovative changing technology. The role of the Federal sector is evolving and the Federal geospatial community will have to be agile and flexible to accommodate and adapt to these changing roles.

The geospatial technology field is at a remarkable point in its evolution, presenting an opportunity to rethink the deployment and use of these resources across the Federal sector, and to enhance our ability to solve problems using geographic information while ultimately saving time and money. This plan provides a new set of goals, objectives, and actions to advance Federal geospatial programs in light of the changing state of geospatial technology and capabilities.

IMPLEMENTING THE NSDI STRATEGIC PLAN

Implementation Strategies and Performance Measurement Approach

The FGDC community will work collaboratively to implement and monitor the implementation of the NSDI Strategic Plan. The FGDC Executive Committee will have the lead responsibility for overseeing and monitoring the implementation of the plan. A designated Federal official, appointed from the FGDC Executive Committee, will serve as the champion for each strategic goal. The champions will report to the FGDC Executive Committee and Steering Committee. The champions will be responsible for overseeing team leaders and action teams, addressing each objective and its supporting actions.

Following adoption of the NSDI Strategic Plan, each action team leader will develop a project plan outlining how the goals and objectives will be achieved. The project plans will include:

- Detailed tasks and milestones
- Timelines
- Responsible parties
- Performance indicators/measures
- Identification of resources available to achieve the objectives

FGDC leadership will assist in identifying and committing resources as well as tracking progress.

The performance measurement approach is critical to the successful implementation of the goals and objectives described in this plan. Performance measures will be based on data (numeric description and results) and will describe whether the identified action is achieving its expected result and if progress is being made toward attaining the goal or objective. With the completion of each project plan, performance measures will be reviewed and evaluated by the champions and Executive Committee as to whether the measure or metric will demonstrate progress toward achieving the objective.

The champions, in collaboration with the Executive Committee, will monitor performance based on review of performance measures and milestones. The FGDC will develop appropriate processes and tools to monitor and track the implementation of the plan. Brief progress reports will be required for each objective on an ongoing basis. Progress will be reported to the FGDC Steering Committee and Executive Committee at each meeting. Updates and adjustments to the plans will be completed as needed.

The implementation of the NSDI Strategic Plan will be aligned with the implementation of the geospatial portfolio management approach described in the Supplemental Guidance to OMB Circular A–16. These initiatives will lead to a more efficient and systematic approach to the development and management of National Geospatial Data Assets.

Collaboration with Non-Federal Partners

While this plan focuses primarily on the Federal role in achieving the shared national vision for the NSDI, the successful implementation of the goals and objectives in the plan will require coordination with a wide range of non-Federal partners and stakeholders. The FGDC will work closely with the NGAC and other partner organizations to develop collaborative approaches to developing and maintain the Nation's critical geospatial infrastructure.

Additional Information

Member Agencies

About the FGDC

The Federal Geographic Data Committee is organized under the authority of OMB Circular A–16 (revised 2002) and Executive Order 12906. The FGDC provides leadership for the NSDI and coordinates the development, use, sharing, and dissemination of the Nation's geospatial data. In accordance with Circular A–16, the FGDC is chaired by the Secretary of the Interior (or designee), and the OMB Deputy Director for Management (or designee) serves as Vice Chair. The FGDC includes the following member agencies:

Federal Communications Commission*	U.S. Department of Commerce
General Services Administration	U.S. Department of Defense
Library of Congress	U.S. Department of Education
National Aeronautics and Space Administration	U.S. Department of Energy
National Archives and Records Administration	U.S. Department of Health and Human Services
National Capital Planning Commission*	U.S. Department of Homeland Security
National Science Foundation	U.S. Department of Housing and Urban Development
Office of Management and Budget	U.S. Department of the Interior
Office of Personnel Management	U.S. Department of Justice
Small Business Administration	U.S. Department of Labor
Smithsonian Institution	U.S. Department of State
Social Security Administration	U.S. Department of Transportation
Tennessee Valley Authority	U.S. Department of the Treasury
U.S. Agency for International Development	U.S. Department of Veterans Affairs
U.S. Army Corps of Engineers*	U.S. Environmental Protection Agency
U.S. Department of Agriculture	U.S. Nuclear Regulatory Commission

*Non-voting members.

National Geospatial Advisory Committee

The National Geospatial Advisory Committee (NGAC) is a Federal Advisory Committee established by the Department of the Interior to provide advice and recommendations related to management of Federal and national geospatial programs, development of the NSDI, and implementation of OMB Circular A–16 and Executive Order 12906. The Committee reviews and comments on geospatial policy and management issues and provides a forum to convey views representative of non-Federal stakeholders in the geospatial community. The NGAC includes representatives from all levels of government, the private sector, nonprofit organizations, and academia. Additional information about the NGAC, including a list of committee members, is available at www.fgdc.gov/ngac.

List of Authorities

Office of Management and Budget, 2000, November 28, Circular A–130 Revised, accessed November 15, 2013, at http://www.whitehouse.gov/omb/circulars_a130_a130trans4/

Office of Management and Budget, 2002, August 19, Circular A–16 Revised, accessed November 15, 2013, at http://www.whitehouse.gov/omb/circulars_a016_rev/

Office of Management and Budget, 2006, March 3, Memorandum M–06–07: Designation of a Senior Agency Official

for Geospatial Information, accessed November 15, 2013, at https://www.fgdc.gov/participation/steering-committee/memos/EOP-3-3-06-Designation-of-SAOGIs_m06-07.pdf

Office of Management and Budget, 2010, November 10, Circular A–16 Supplemental Guidance, accessed November 15, 2013, at http://www.whitehouse.gov/sites/default/files/omb/memoranda/2011/m11-03.pdf

Office of Management and Budget, 2012, May 2, Federal Information Technology Shared Services Strategy, accessed November 15, 2013, at http://www.whitehouse.gov/sites/default/files/omb/assets/egov_docs/shared_services_strategy.pdf

Office of Management and Budget, 2013, May 9, Open Data Policy—Managing Information as an Asset, accessed November 15, 2013, at http://www.whitehouse.gov/ sites/default/ files/omb/memoranda/2013/m-13-13.pdf

Public Law 104–106, 1996, February 10, National Defense Authorization Act for Fiscal Year 1996 (includes

Clinger-Cohen Act), accessed November 15, 2013, at http://www.gpo.gov/fdsys/pkg/PLAW-104publ106/pdf/PLAW104publ106.pdf

Public Law 107–347, 2002, December 17, E-Government Act of 2002, accessed November 15, 2013, at http://www.gpo.gov/fdsys/pkg/PLAW-107publ347/content-detail.html

The White House, 1994, April 11, Executive Order 12906: Coordinating Geographic Data Acquisition and Access: The National Spatial Data Infrastructure, accessed November 15, 2013, at http://www.archives.gov/federal-register/ executive-orders/pdf/12906.pdf

Additional References

Oxera Consulting Ltd, 2013, January, What is the Economic Impact of Geo Services?, accessed November 15, 2013, at http://www.oxera.com/Oxera/media/Oxera/downloads/reports/What-is-the-economic-impact-of-Geo-services_1. pdf?ext=.pdf

The Boston Consulting Group, 2012, December, Putting the U.S. Geospatial Industry on the Map, accessed November 15, 2013, at http://www.ncge.org/files/documents/US-FullReport.pdf

U.S. Government Accountability Office, 2012, November 26, Geospatial Information - OMB and Agencies Need to Make Coordination a Priority to Reduce Duplication (GAO-13-94), accessed November 15, 2013, at http://gao.gov/ products/GAO-13-94

GLOSSARY OF ABBREVIATIONS AND TERMS

3D	Three dimensional
4D	Four dimensional
BCG	Boston Consulting Group
CIO	Chief Information Officer
CUI	Controlled, unclassified information
DOI	Department of the Interior
FGDC	Federal Geographic Data Committee
FY	Fiscal year
GAO	Government Accountability Office
GPS	Global Positioning System
GSDI	Global Spatial Data Infrastructure
IT	Information technology
MOOC	Massive open online courses
NGAC	National Geospatial Advisory Committee
NGDA	National Geospatial Data Assets
NSDI	National Spatial Data Infrastructure
OMB	Office of Management and Budget
Open Standards	See OMB Open Data Policy—Managing Information as an Asset (http://www.whitehouse.gov/sites/default/files/omb/memoranda/2013/m-13-13.pdf)
SDI	Spatial Data Infrastructure

UN-GGIM	United Nations Initiative on Global Geospatial Information Management
USGEO	U.S. Group on Earth Observations

INDEX

D

E

F

G

H

I

J

L

M

N

O

P

Q

R

S

T

U